# REGIONAL CLIMATE CHANGE EFFECTS:
# USEFUL INFORMATION FOR TRANSPORTATION AGENCIES

Prepared for:
Office of Planning, Environment and Realty
Office of Infrastructure
Federal Highway Administration, U.S. DOT

Prepared by:
ICF International
1725 Eye Street, NW
Washington, DC 20006

May 10, 2010

Contract No. DTFH61-05-D-00019; TOPR No. EV0101

U.S. Department
of Transportation

**Federal Highway
Administration**

## Acknowledgments

FHWA would like to thank the individuals listed below who provided guidance and feedback on the climate effects information included in this report. Their insights and information were critical to ensuring the scientific credibility of the data presented here. FHWA looks forward to continued dialogue regarding the nature of the climate change data that will be most useful to transportation practitioners in their efforts to adapt the transportation system to a changing climate.

- Virginia R. Burkett, U.S. Geological Survey
- Arthur T. Degaetano, Cornell University
- Katharine Hayhoe, Texas Tech University
- David H. Levinson, National Oceanic and Atmospheric Administration
- A. David McGuire, U.S. Geological Survey, University of Alaska
- Philip W. Mote, Oregon Climate Change Research Institute
- Thomas C. Peterson, National Oceanic and Atmospheric Administration
- Kelly T. Redmond, Western Regional Climate Center
- David A. Robinson, Rutgers University, New Jersey
- E. Robert Thieler, U.S. Geological Survey
- Michael F. Wehner, Lawrence Berkeley Laboratory
- Donald J. Wuebbles, University of Illinois

FHWA would further like to thank Thomas C. Peterson, Jay H. Lawrimore, Michael Wehner, and Katharine Hayhoe for providing USGCRP climate projection data and figures that are used extensively in this report. Though every effort was made to ensure accuracy, FHWA and the authors are responsible for the information in this report.

## Table of Contents

Executive Summary ................................................................................................................. i

1 Introduction: Why Should the Transportation Community be Concerned with Global Climate Change? .................................................................................................................. 1
    1.1 Observed Changes ................................................................................................ 3
    1.2 Potential Future Impacts ....................................................................................... 5

2 Methodology Overview ................................................................................................ 12

3 Projected Climate Change by Geographic Region ....................................................... 18
    3.1 National projections ............................................................................................ 18
    3.2 Regional Summaries ........................................................................................... 28
    3.3 Northeast ............................................................................................................. 30
    3.4 Southeast ............................................................................................................. 36
    3.5 Midwest ............................................................................................................... 41
    3.6 Great Plains ......................................................................................................... 46
    3.7 Southwest ............................................................................................................ 49
    3.8 Pacific Northwest ................................................................................................ 54
    3.9 Alaska ................................................................................................................. 59
    3.10 Hawaii ................................................................................................................. 63
    3.11 Puerto Rico ......................................................................................................... 66

4 Future Work .................................................................................................................. 69
    4.1 Improvements to current projections .................................................................. 69
    4.2 Additional climate variables or enhanced techniques ........................................ 70

5 Acronyms and Glossary ................................................................................................ 71

6 References ..................................................................................................................... 79

Appendix A. Detailed Methodology

Appendix B. Regional Maps

Appendix C. Climate Change Effects Typology Matrix

## Executive Summary

*"Climate affects the design, construction, safety, operations, and maintenance of transportation infrastructure and systems. The prospect of a changing climate raises critical questions regarding how alterations in temperature, precipitation, storm events, and other aspects of the climate could affect the nation's roads, airports, rail, transit systems, pipelines, ports, and waterways." CCSP 2008a*

The changing climate poses serious challenges to the transportation community, given the community's need to watch over transportation systems and infrastructure designed to last decades or longer. Transportation functions tied to construction, operations, maintenance, and planning should be grounded in an understanding of the environment expected to support transportation facilities. Decisions therefore need to be informed by an understanding of potential future changes in climate. The understanding of climate change science and the ability to model future change continues to advance, resulting in more precise estimates of future changes in climate. However, the practitioner can be overwhelmed by the sheer volume of information, including the ensemble of models employed, the variety of emissions scenarios used to drive the modeling results, the spatial resolution of the projected climate effects, and other parameters.

The purpose of this report is to provide the transportation community (including highway engineers, planners, NEPA practitioners) with digestible, transparent, regional information on projected climate change effects that are most relevant to the U.S. highway system. This information is designed to inform assessments of the risks and vulnerabilities facing the current transportation system, and can inform planning and project development activities.

Why should the transportation community care about this information? The impacts of climate change can include weakened bridges and road beds, temporarily or permanently flooded roads, damaged pavements, and changes in road weather that can affect safety and economic activity. Understanding and proactively addressing the potential impacts of climate change can help avoid the potential damage, disruption in service, and safety concerns that climate change may cause.

Climate change information is provided in this report by U.S. region, by time horizon, and by climate variable or "climate effect" (i.e., changes in temperature, precipitation, storm activity, and sea level). The multi-state regions are identical to those included in the U.S. Global Change Research Program (USGCRP) climate impact analyses (USGCRP 2000, 2009). Three time horizons were chosen for each region: *near-term* (2010-2040), *mid-century* (2040-2070), and *end-of-century* (2070-2100).

FHWA's initial research efforts attempted to capture regional or sub-regional projections from all publicly available, peer-reviewed studies for these climate effects. During the course of this research, FHWA consulted with a range of nationally recognized climate scientists to ask for their insights and recommendations regarding the most credible regional projections for use by State DOTs and local transportation agencies. As a result, FHWA obtained key data sets not previously published in their entirety, including data compiled from the CMIP3 database of climate model integrations by Michael Wehner of the Lawrence Berkeley National Laboratory for the USGCRP's *Global Climate Change Impacts in the United States* (2009) report. These and other data sets were further evaluated and scrutinized, and subsequently vetted with a set of regional climate experts.

The results of this research provide a two-part resource to transportation practitioners:

1. Projected climate change effects in narrative form, supported by tables and maps, for variables of interest, for specific time horizons, representing consistent emission scenarios and a consistent presentation of uncertainty.

2. Detailed climate change projections for regions and, to the extent possible, sub-regional areas (e.g., cities, states) for all time horizons, uncertainty levels, emission scenarios, and time horizons housed in the Climate Change Effects Typology Matrix (Appendix C).

The process used to create this report and the Climate Change Effects Typology Matrix has resulted in a unique quantitative and qualitative regional analysis of the best available climate projections. It also serves as a platform for discussions between transportation officials and climate scientists. Looking forward, the approach used here and the relationships forged in creating this set of climate information will assist future efforts to refine and disseminate information on climate effects to transportation practitioners.

Some of the findings of this report are briefly outlined below. These illustrative findings are by region and mostly for *mid-century* (i.e., showing projected effects in 2040 to 2070 relative to a 1961 to 1979 baseline), unless otherwise noted, and are based on averages from a multi-model ensemble for a low emission scenario (B1) as well as a moderately high ("business as usual") emission scenario (A2) (USGCRP 2009)[1]:

- o The Northeast is projected to become substantially warmer with an annual mean increase of 3.8 to 4.8°F, and wetter, particularly during the winter months. The duration of heavy precipitation events (i.e., more than 2 inches per day) is projected to increase along with the average amount of rain falling within a given rainy day.

- o The Southeast is projected to undergo a 3.2 to 4.0°F increase in annual mean temperature, with greater warming and reduced precipitation during the summer and fall months.

- o The Midwest is projected to experience an annual mean temperature increase of 4.0 to 5.0° F, with much wetter winters and springs and drier summers.

---

[1] For descriptions of emissions scenarios, see Appendix A.

- The Great Plains' annual mean temperature is projected to increase by 3.8 to 4.7°F, with wetter winters and drier summers.

- The Southwest is projected to experience an annual mean warming of 3.6 to 4.5°F, with summers and falls experiencing the greatest increases. The Southwest's winters are projected to be somewhat wetter, while the spring months, in particular, are projected to be substantially drier.

- The Pacific Northwest is projected to experience an annual mean temperature increase of 3.6 to 4.3°F, with the greatest warming and greatest reductions in precipitation projected for the summer months.

- Alaska is projected to experience the greatest warming of any U.S. region, with increases in annual mean temperature of 4.3°F, and the greatest warming expected during the winter months. Precipitation increases are projected year round, ranging from 9 to 17%, depending on the season.

- Annual mean temperature on the Hawaiian Islands is projected to increase by 2.7 to 3.3°F. Hawaii's precipitation is projected to increase during the fall months while the other seasons are projected to experience a decrease.

Recent estimates of global average sea-level rise by the end of the century range from 7 to 79 inches (IPCC 2007a; Rahmstorf 2007; Grinsted et al. 2009; Rohling et al. 2008; Pfeffer et al. 2008). Most coastal regions in the contiguous United States are expected to experience sea-level rise of this general magnitude. However, relative sea level will rise more than the global average in regions experiencing a greater rate of subsidence of coastal land (e.g., the Gulf Coast), whereas in fewer areas (e.g., parts of Alaska) local uplift will dominate and relative sea level will rise less than the global average. In addition to projected changes in vertical motion, local sea-level rise may also be affected by such factors as local changes in ocean circulation, ocean density, gravitational effects, sedimentation, and erosion.

Assessing the potential harm related to these climate effects allows highway planners to identify and address vulnerabilities. Many of the risks from climate change come from an increased exposure to weather and climate extremes. Since the highway system is engineered to withstand the historically expected range of weather stressors, small changes in average climate are not expected to cause significant impacts. However, because future climate change is projected to transcend the bounds of historic experience, it is likely to expose vulnerabilities. Impacts could include abrupt and unanticipated disruptions to the system (such as a road washing out), or more gradual disruption (such as an increased need for road maintenance).

In particular, extreme heat days, heavy precipitation events, high wind, and storm surge all pose significant risks to the highway system. For example, extreme heat causes thermal expansion on bridge joints and paved surfaces, which can result in structural degradation. Heavy precipitation events can cause flooding or mudslides that block and damage roads. High winds during severe storms can damage street lights, signs, and overhead cables. Storm surge can cause erosion of the road base and bridge supports. At the same time, climate change can reduce exposure to other risks, particularly those related to cold weather extremes. Decision makers may not wish to respond to every potential climate risk, but identifying those risks will allow them to anticipate potential disruptions and prioritize their responses.

The information in this report can help decision makers begin to address the challenges posed by climate change. It fills an important gap by providing the transportation community with information on climate change and the range of future changes in a usable format. It provides the most up-to-date information available, and is the place to start when seeking to understand how climate change may affect transportation systems and infrastructure. At the same time, this report does not answer every question on future climate change effects; research continues to progress on improving techniques for projecting and assessing climate effects and understanding extreme weather events. In the coming years, model simulations of the effect of changes in greenhouse gas concentrations on the climate will improve, and downscaling techniques that provide finer-scale climate projections will continue to evolve.

# 1 Introduction: Why Should the Transportation Community be Concerned with Global Climate Change?

The changing climate poses serious challenges to the transportation community, given our need to watch over transportation systems and infrastructure designed to last decades or longer. Transportation functions tied to construction, operations, maintenance, and planning need to be based on an understanding of the environment expected to support transportation facilities. Decisions therefore need to be informed by an understanding of potential future changes in climate.

Changes in the climate over the recent past have been documented by researchers, including changes in temperature, precipitation, storm activity, sea level, and wind speeds. These effects can in turn lead to impacts on transportation, such as weakened bridges and road beds, permanently flooded roads, damaged pavements, and changes in road weather that can affect safety (both positively and negatively) and economic activity. Understanding and proactively addressing the potential impacts of climate change can help avoid the potential damage, disruption in service, and safety concerns that climate change may cause.

FHWA is releasing this report in order to provide the transportation community (i.e., highway and bridge engineers, planners, NEPA practitioners, etc.) with transparent and reliable information on projected climate change effects that are most relevant to the U.S. highway system to the extent that such information was available through the summer of 2009.

This report synthesizes available information to present a picture of how the climate might change over the near term (2010-2040), by mid-century (2040-2070), and at the end of this century (2070-2100) for the country divided into nine regions, and it summarizes the current understanding of these projected effects primarily at the regional level. The nine regions match those used in the U.S. Global Change Research Program's *Global Climate Change Impacts in the United States* (2009) report.

There are several objectives in preparing this information. It is designed to help transportation practitioners better understand what climate change means for transportation, generally, and it is also intended to be considered where appropriate in state and metropolitan planning and project development. For example, information on changes in seasonal precipitation can inform analysis of stream flow and scour. Projections of changes in summer temperatures can inform decisions tied to infrastructure design and maintenance.

On the other hand, it represents an initial step in outlining the type and range of potential future climate change effects facing transportation in the United States, and is meant to be a starting point in understanding how climate change may affect transportation. It does not provide all the information needed. The science of projecting future climate is evolving, and in time more and better information should become available.

This report focuses on region-scale projections for temperature and precipitation, which were assembled based on the results of global climate models. The data cited were developed in support of the USGCRP document, but have not previously been published in their entirety. This information exists for all regions of the country. Additionally, the report includes the findings of a literature review in the Climate Change Effects Typology Matrix (Appendix C). The Typology collects relevant projections of climate change effects included in other reports and from peer-reviewed literature. Some of this information is also included in the regional sections contained in chapter 3. This report also includes results that were statistically downscaled from the results of climate models to the sub-regional level. This information, which is presented in several figures in Chapter 3 and Appendix B, is limited to the contiguous 48 states. The downscaling of climate data is an area of continued development, with new techniques likely to become available in the coming years.

While much of the information provided in this report applies to a region as a whole, rates of change in a given location may or may not match projections for the larger region. Thus, transportation agencies will need to work with environmental and engineering staffs to determine what these regional projections mean for their local or state level planning, programming, and maintenance efforts. The information in this report can provide insights into the range of changes and trends that may influence the state, local, or project level.

This report provides a first step in understanding the impacts that climate change may have on existing and future transportation infrastructure and operations. Additional work will be needed to understand how best to incorporate this information into project- and system-level planning, to translate this information into specific impacts on transportation, and to assess what if any modifications are necessary to the transportation network to ensure its long-term ability to provide access and mobility for people and goods. FHWA will undertake further work to develop tools, methods, and ultimately guidance that can be used to apply climate change information to decision-making. We will learn from further research (including our current work to develop preliminary tools to assess the vulnerability and risk of transportation to the effects of climate change) and the efforts of the transportation community, including State and local practitioners, on how best to apply this information. Each program area—asset management, metropolitan and statewide planning, project development, operations, safety, infrastructure, bridge, pavements, etc.—may apply this information differently based on its specific needs, local context, or other factors. As we gain more experience using this information, and the science is updated and refined, we will provide additional guidance.

This chapter briefly examines climate variability experienced through the end of the 20th century, focusing on the national level. It also summarizes a range of potential climate impacts on the highway system that can result from changes in climate, and introduces concepts such as risk and vulnerability. We have included this information early in the report—before discussing the regional climate effects—to illustrate the importance of examining the effects of climate change. Chapter 2 provides a brief overview of both the methodology used to assemble the climate data for this report and the consultation process with national and regional climate experts. (This methodology is treated in more detail in Appendix A.) Chapter 3 summarizes and discusses the available data and literature of projected climate change effects, focusing on the national and regional levels. This information serves as the foundation of this report; with the rest of the information providing context. Appendix B illustrates the Chapter 3 tables describing regional climate projections of annual and seasonal temperature and seasonal precipitation. Chapter 4 discusses needed future work, and identifies climate projections that are currently unavailable. Chapter 5 includes a glossary of terms used in the report. Chapter 6 provides the list of references.

Appendix C provides the Climate Change Effects Typology Matrix and will be posted online. It provides one-line summaries of the results of each current collected study or set of modeled data for each of the main climate effects, by region, and time frame. It was assembled while locating relevant information for this report. (Chapter 2 and Appendix A provide a description of the methodology used in developing the Climate Change Effects Typology Matrix.)

## 1.1 Observed Changes

This section provides a brief review of national-scale changes in temperature, precipitation and storm activity, and sea level observed in recent decades. This information is drawn largely from the IPCC (2007a), National Science and Technology Council (2008), and USGCRP (2009) reports, and is based on satellite measurements and data from thousands of weather stations, ships, and buoys around the world carefully compiled by independent research groups.[2]

### 1.1.1 Temperature

Over the 20$^{th}$ century, the Earth's annual average temperature has increased by approximately 1.3 ± 0.32°F (0.74 ± 0.18°C) (IPCC 2007a). The winter and spring seasons in the Northern Hemisphere have experienced the greatest degree of warming, with the United States experiencing a warming of near 0.58°F per decade over the past few decades (National Science and Technology Council 2008). As the frequency of heat waves has increased, the number of unusually cold days has decreased (National Science and Technology Council 2008; USGCRP 2009). However, over the past few decades, the diurnal temperature difference has not changed, with day and night temperatures rising at similar rates (USGCRP 2009).

The impact of this warming on the natural system has already been well documented. For example, the area of Arctic sea ice has shrunk at a rate of about 2.7 percent per decade, with the summer months experiencing even greater reductions of 7.4 percent per decade. This is a result of warming in the Arctic that is twice the average warming in the United States (USGCRP 2009). In the continental United States and Alaska (the middle and high latitudes), shifts in phenology—the timing of life cycles events of plants and animals—have been noted. These phenological changes include an increase in the growing season

---

[2] Differences caused by changes in instruments, measurement times and locations, etc. were taken into account during data processing in the reports referenced.

of approximately 2 weeks since 1950, and earlier annual occurrences of plant flowering and animal spring migration (USGCRP 2009).

### 1.1.2 Precipitation and Storm Events

Over the 20th century, the total average annual precipitation for the contiguous United States increased by 6 percent (National Science and Technology Council 2008). During the second half of the 20th century, some regions across the United States have experienced increases in drought severity and duration as temperatures have risen (USGCRP 2009). The United States has experienced extreme drought events in the past, such as during the mid-1930s when portions of the Great Plains became known as the "dust bowl" due to wind erosion brought on by several years of drought and compounded by the replacement of moisture-retaining natural vegetation with crops. Since the 1950s, parts of the Southeast and West have experienced an increase in drought conditions, while the Midwest and Great Plains have experienced a reduction.

During the last three decades of the 20th century, the eastern United States experienced an increase in heavy precipitation events[3] (National Science and Technology Council 2008), and an increase in the proportion of total annual precipitation that falls during heavy precipitation events.

While recent research indicates there is some likelihood that the number of tropical storms and hurricanes each year in the North Atlantic has increased over the past 100 years (National Science and Technology Council 2008), the number of hurricanes that make landfall has stayed relatively constant (USGCRP 2009). The intensity of the strongest hurricanes is also likely to have increased in this region: sea surface warmth is a strong contributor to tropical storm development, and climate change is considered to have contributed to the increase of sea surface temperatures in the North Atlantic and Northwest Pacific hurricane formation regions (CCSP 2008b). However, it is unclear if and how other tropical storm development factors—such as temperature and moisture profiles, wind shear, or near-surface ocean temperature stratification—have changed. The trend is further complicated by multi-decadal variability and data-quality issues. For smaller-scale phenomena such as tornadoes, hail, lightening, and dust storms, the IPCC (2007a) concluded that there is insufficient evidence to determine the associated trends.

### 1.1.3 Sea-Level Rise

Over the 20th century, global average sea level has risen by 6.7 inches (0.17 meters) (IPCC 2007a). Figure 1 below demonstrates how sea-level rise varies regionally across the United States. These differences are due primarily to differences in vertical land motions (USGCRP 2009). Relative sea level is rising 0.8 to 1.2 inches per decade along most of the Atlantic and Gulf Coasts, with a few inches per decade occurring along the Louisiana Coast due to land subsidence (National Science and Technology Council 2008). Other regions of the country, such as specific coastline locations for Alaska, are

---

[3] Heavy precipitation events are defined by the referenced reports as an event with at least 2 inches of precipitation per day. For purposes of this report, heavy precipitation events constitute a storm event; however, given the limitation of the information provided, no discussion of the type of storm, the associated phenomena such as winds, nor the related stressors such as flooding can be determined.

experiencing land uplift and a corresponding relative sea level decline of a few inches per decade (National Science and Technology Council 2008). Sea-level rise increases the risk of impact from storm surge and waves farther inland, causing shoreline erosion and local damage.

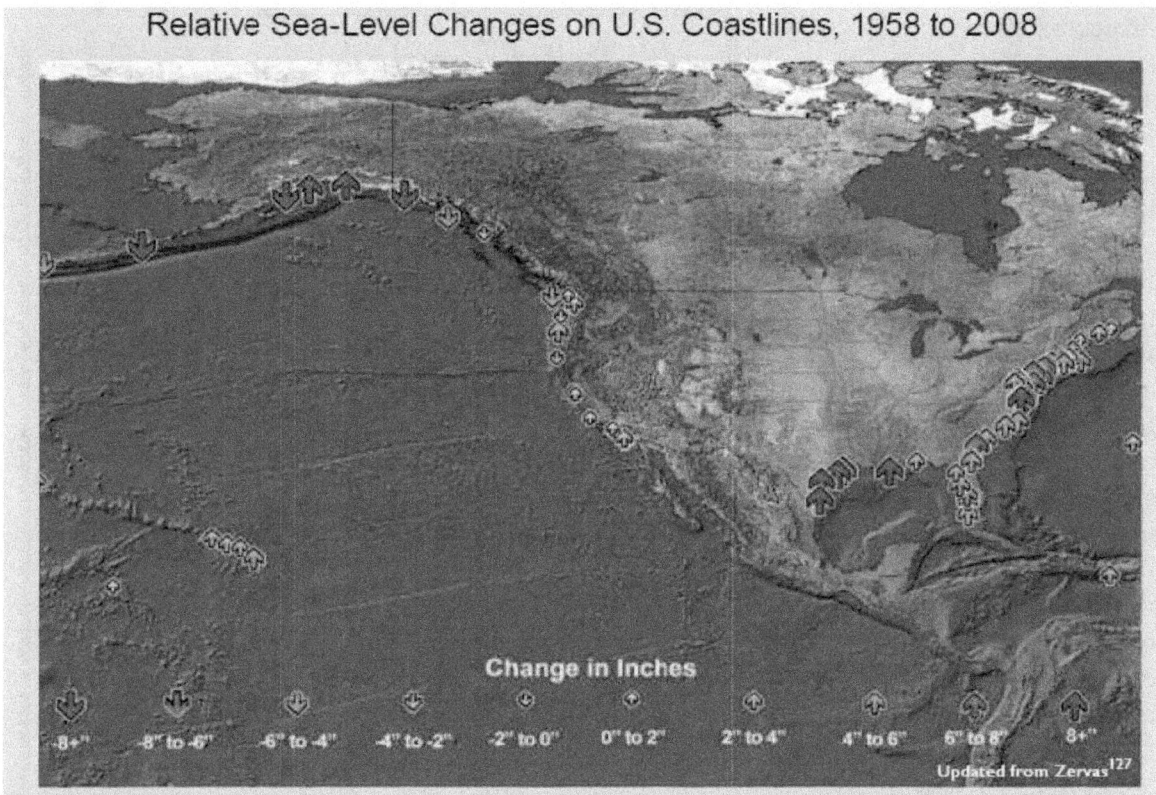

Figure 1: Observed changes in relative sea level from 1958 to 2008 for locations on the U.S. coast. Some areas along the Atlantic and Gulf coasts saw increases greater than 8 inches over the past 50 years. (Source: Zervas 2001)

## 1.2 Potential Future Impacts

Assessing the potential harm of climate stressors allows vulnerabilities to be addressed before they become problems. For example, knowing that a road used during emergency evacuations will be at risk for failure due to erosion allows decision makers to decide what to do *before* the road washes out. Transportation decision makers may decide to take measures to prevent the road from washing out or find another route. They may also decide that the costs of action are too great, and the risks are too low (or the likelihood of damage is too low), to justify any action. However, in order to make such decisions, an assessment of the climate-related risks is necessary.

## 1.2.1 Connecting Climate Changes to Impacts

Projected changes in temperature, precipitation, storm activity, wind, and sea level indicate the magnitude of the stressors to which highway infrastructure could be exposed in the future. However, these effects do not in and of themselves indicate what the ultimate consequences to the highway infrastructure will be. From a highway operations standpoint, does it actually matter that temperatures or rainfall might increase? And would these effects translate to beneficial or adverse impacts on the highway system? The answers depend on many factors, such as the severity of the climate stressor, the engineering and design characteristics of the structures, geographic and geologic characteristics, and operations and maintenance activities. It is therefore important to consider all of these factors when assessing the potential impacts of climate change.

> **Vulnerability** describes how susceptible a system is to the adverse effects of climate change (IPCC 2007b).
>
> **Vulnerability Factors** include the age of the infrastructure element, condition/integrity of the infrastructure element, proximity to other infrastructure elements/concentrations, and the level of service (CCSP 2008).
>
> **Exposure** is the degree to which a system comes into contact with climate conditions or specific climate impacts (CIG 2007), and the probability, or likelihood, that this stress will affect transportation infrastructure (CCSP 2008a).
>
> **Risk** characterizes both the probability of the event occurring and the consequence of the event (Snover et al. 2007; NZCCO 2004).
>
> **Potential Climate Impacts** describes how projected climate effects may affect the highway system through current or newly introduced system exposures or sensitivities. It should be noted that the uncertainties associated with projecting the impacts continue to apply when considering risks (which also have the additional uncertainty of consequences).

While consideration of so many variables may seem daunting, it is important to remember that highway structures are already being exposed to many climate- or weather-related stressors. Climate change can exacerbate (or lessen) these same stressors. Highway infrastructure is already exposed to (and, to a certain extent, designed to withstand) the elements. An area that rarely experiences severe storms is unlikely to suddenly experience frequent hurricanes. Instead, an area already exposed to severe storms may experience more severe or more frequent storms. Since infrastructure is designed to withstand locally expected climate stressors of the magnitude and frequency that have historically been experienced, the risks from climate change can come from an amplification of existing stressors (NRC 2008).

In some cases, climate changes will be sufficient to push some aspect of a transportation system over a certain threshold. For example, rising sea levels can introduce erosion problems to areas that previously had limited exposure to those threats, and thereby undermine roadways that had previously been unaffected by erosion. For inland areas, falling lake levels can lead to erosion of bedrock due to wave impacts, causing roads to fail.

Many of the risks from climate change come from an increased exposure to extremes in weather and climate. One example is a projected increase in the number of days with extremely high temperatures, which cause more stress than simply an increase in the average temperature. Most infrastructures are engineered to withstand a normal or expected range of climate or weather stressors, and small changes in average climate won't have significant impacts on the structures themselves (NRC 2008). However, as the climate changes, the expected range of stressors may not accurately reflect actual exposure. For example, bridges are often designed to withstand a certain level of flooding (such as the height of a flood that normally has a 1% likelihood of occurring in a year, i.e., a 1 in 100 year flood event), a probability that is typically based on historical data. Under some climate change scenarios, a flood level that has a true 1% annual likelihood of occurring might actually be much more severe than the historical 1% return rate flood.

It is also important to note that not all climate change impacts are negative. In some areas, climate change could reduce the frequency, duration, and/or severity of some cold-weather extremes, for example. In these cases, shorter winters may lead to longer thaw seasons, or higher temperatures could allow an extension in the work season—both of which are potential benefits. But generally, it is useful to keep in mind that any deviation from the climate for which infrastructure has been designed is likely to cause negative impacts (CCSP 2008b; USGCRP 2009).

### 1.2.2 Potential Impacts to Highways

Climate change impacts on highway infrastructure can be sudden and severe, or may occur more gradually. Floods and erosion can completely and abruptly shut down a road. In contrast, an increase in the frequency and severity of extremely high temperatures can lead to pavement deterioration and rutting. These more gradual problems can be predicted and addressed through additional maintenance, but they are still costly and disruptive to traffic flow. This difference is important to remember when assessing priorities in climate adaptation. Are decision makers concerned primarily with sudden and severe highway system failures, or more generally with minimizing costs and travel disruptions?

When considering potential impacts on the highway system, it is also important to keep in mind the interconnectivity of the highway system. When looking at a specific geographic area, it can be tempting to dismiss climate stressors as not being relevant to that area. But disruptions in highway systems in one location can affect the timely delivery of goods in another location; they can also disrupt passenger travel. Additionally, areas most affected by climate change might ultimately draw larger amounts of money to repair damages and reduce future damages—diverting resources that could have been applied elsewhere.

Highway infrastructure and operations/maintenance are affected by three key categories of climate change effects: changes in temperature, changes in precipitation and storm events, and sea-level rise. This section continues by providing some example impacts that these climate effects can have on highways. The information provided in the tables below is based on a collection of existing reports and is provided for this discussion; however, the tables do not represent all plausible potential impacts. Actual impacts would depend on local conditions as well as the severity of climate change effects.

### 1.2.3 Changes in Temperatures

Small changes in average temperatures will not greatly affect highway systems; however, by the end of the century some regions are projected to experience significant warming that will affect highway planning, construction, and operations. Significant impacts may also occur with increases in both the intensity (how high the high temperature is) and duration of very hot periods, or decreases in the intensity and duration of periods with very cold days. Extreme temperatures can cause both structural damage to highway assets and challenges to the use and maintenance of the roads. As noted in Section 3, the number of extreme heat days is project to increase in all regions of the continental United States. Meanwhile, the shortening of the winter season could provide some benefits in terms of lengthening the construction season and reducing snow/ice removal costs, particularly in the more northern regions. However, the northern states are projected to sustain increases in freeze-thaw conditions,[4] potentially increasing the occurrence of frost heaves and potholes on road and bridge surfaces (USGCRP 2009). Table 1-1 provides a summary of how changes in temperature may affect highway infrastructure and operations.

| Climate Effects | Impacts on Infrastructure and Operations |
|---|---|
| Increases in very hot days and heat waves (higher high temperatures, increased duration of heat waves) | • Increased thermal expansion of bridge joints and paved surfaces, causing possible degradation<br>• Concerns regarding pavement integrity, traffic-related rutting and migration of liquid asphalt, greater need for maintenance of roads and pavement<br>• Maintenance and construction costs for roads and bridges; stress on bridge integrity due to temperature expansion of concrete joints, steel, asphalt, protective cladding, coats, and sealants<br>• Asphalt degradation, resulting in possible short-term loss of public access or increased congestion of sections of road and highway during repair and replacement<br>• Limits on periods of construction activity, and more nighttime work<br>• Vehicle overheating and tire degradation |
| Decreases in very cold days | • Regional changes in snow and ice removal costs, environmental impacts from salt and chemical use<br>• Changes in pavement designs<br>• Fewer cold-related restrictions for maintenance workers |
| Later onset of seasonal freeze and earlier onset of seasonal thaw | • Changes in seasonal weight restrictions<br>• Changes in seasonal fuel requirements |

---

[4] Freeze-thaw conditions refer to the number of days when the maximum temperature is greater than freezing and the minimum temperature is below freezing.

|  | • Improved mobility and safety associated with a reduction in winter weather<br>• Longer construction season in colder areas |
|---|---|

Table 1-1: Impacts of temperature on highway operations and infrastructure. Sources: NRC (2008), CCSP (2008a), CSIRO (2006), Department of Transport (U.K.) (2004)

### 1.2.4 Changes in Precipitation and Storm Events

Increased rains can cause disruptions in the use of highways (mainly due to flooding), as well as structural damage. In some areas, however, drought conditions are expected to increase, which can introduce other threats to the highway system. For example, summer precipitation is expected to decrease for most regions (see Section 3), which could lead to isolated pockets of increased drought-related impacts. Winter precipitation for most regions is projected to increase, which *could* result in additional snow and ice removal costs (particularly for the Northeast Great Plains and Midwest states).[5] However, in many cases, those impacts are generally projected to be less in areas and at times of year that are now just below freezing but projected to warm above freezing in the future.

Storms such as tropical cyclones, thunderstorms, and extratropical cyclones can cause sudden, dramatic, and costly disruptions to the highway systems. Use of the highways can be disrupted as a result of flooding or structural failures. Severe winds and rains can also cause significant damage to structures. Table 1-2 provides a summary of how changes in severe storm intensity may affect highway infrastructure and operations.

| Climate Effects | Impacts on Infrastructure and Operations |
|---|---|
| Increases in intense precipitation events | • Increases in weather-related delays and traffic disruptions<br>• Increased flooding of evacuation routes<br>• Increases in flooding of roadways and tunnels<br>• Increases in road washout, landslides, and mudslides that damage roadways<br>• Drainage systems likely to be overloaded more frequently and severely, causing backups and street flooding<br>• Areas where flooding is already common will face more frequent and severe problems<br>• If soil moisture levels become too high, structural integrity of roads, bridges, and tunnels (especially where they are already under stress) could be compromised<br>• Standing water may have adverse effects on road base<br>• Increased peak streamflow could affect scour rates and influence the size requirement for bridges and culverts |
| Increases in drought conditions | • Increased susceptibility to wildfires, causing road closures due to fire threat or reduced visibility<br>• Increased risk of mudslides in areas deforested by wildfires |
| Changes in seasonal | • Benefits for safety and reduced interruptions if frozen precipitation |

---

[5] Personal communication with Michael Wehner of the Lawrence Berkeley National Laboratory.

| precipitation and stream flow patterns | shifts to rainfall |
|---|---|
| | • Increased risk of floods, landslides, gradual failures and damage to roads if precipitation changes from snow to rain in winter and spring thaws |
| | • Increased variation in wet/dry spells and decrease in available moisture may cause road foundations to degrade |
| | • Degradation, failure, and replacement of road structures due to increases in ground and foundation movement, shrinkage and changes in groundwater |
| | • Increased maintenance and replacement costs of road infrastructure |
| | • Short-term loss of public access or increased congestion to sections of road and highway |
| | • Changes in access to floodplains during construction season and mobilization periods |
| | • Changes in wetland location and the associated natural protective services that wetlands offer to infrastructure |
| Increases in coastal storm intensity (leading to higher storm surges/wave heights, increased flooding, stronger winds) | • More frequent and potentially more extensive emergency evacuations |
| | • More debris on roads, interrupting travel and shipping |
| | • Bridges, signs, overhead cables and other tall structures are at risk from increased wind speeds |
| | • Increased storm surge and wave impacts on bridge structures |
| | • Decreased expected lifetime of highways exposed to storm surge |
| | • Risk of immediate flooding, damage caused by force of water, and secondary damage caused by collisions with debris |
| | • Erosion of land supporting coastal infrastructure and coastal highways |
| | • Damage to signs, lighting fixtures, and supports |
| | • Reduced drainage rate of low-lying land after rainfall and flooding events |
| | • Damage to infrastructure caused by the loss of coastal wetlands and barrier islands |

Table 1-2: Impacts of precipitation on highway operations and infrastructure. Sources: NRC (2008), CCSP (2008a), CSIRO (2006), personal communication with E. Robert Thieler.

### 1.2.5 Sea-Level Rise

Rising sea levels can permanently inundate coastal roads and cause damaging erosion. Higher sea levels can exacerbate the effects of storm surge, causing storm surges to reach greater heights and further inland, possibly inflicting additional damages on structures. Sea-level rise presents significant risks to many regions, particularly those where land is subsiding. Table 1-3 provides a summary of how changes in sea-level rise may affect highway infrastructure and operations (as changes in storm surge are a function of both storm activity and sea-level rise, the impacts provided in Table 1-3 overlap somewhat with those at the end of Table 1-2).

| Climate Effects | Impacts on Infrastructure and Operations |
|---|---|
| Rising sea levels (exacerbating effect of higher storm surge, increased salinity of rivers and estuaries, flooding) | • Exposes more areas to effects of storm surge/wave action, causing more frequent interruptions to coastal and low-lying roadway travel<br>• Amplifies effect of storm surge, causing more severe storm surges requiring evacuation<br>• Permanent inundation of roads or low-lying feeder roads in coastal areas. Reduces route options/redundancy<br>• More frequent or severe flooding of underground tunnels and low-lying infrastructure, requiring increased pumping activity<br>• As the sea-level rises, the coastline will change and highways that were not previously at risk to storm surge and wave damage may be exposed in the future<br>• Erosion of road base and bridge supports/scour<br>• Highway embankments at risk of subsidence/heave<br>• Reduced clearance (including freeboard) under bridges<br>• Increased maintenance and replacement costs of tunnel infrastructure |

**Table 1-3: Impacts of sea-level rise on operations and highway infrastructure. Sources: NRC (2008), CCSP (2008a), CSIRO (2006), ICF (2007)**

## 2    Methodology Overview

The projections of climate change effects presented in this report were developed through a systematic process initiated by FHWA in the summer of 2009. This section describes the key elements of FHWA's methodology; Appendix A provides more details on the approach, which provides climate change information by U.S. region, by time horizon, and by climate variable. The regions are identical to those used in the U.S. Global Change Research Program (USGCRP) climate impact analyses (USGCRP 2000, 2009). Three time horizons were chosen for each region: *near-term* (2010-2040), *mid-century* (2040-2070), and *end-of-century* (2070-2100). Projected changes in climate are reported for temperature, precipitation, storm events, and sea-level rise.

As described below, initial research efforts attempted to capture regional or sub-regional projections from all publicly available, peer-reviewed studies for these climate effects. During the course of this research, FHWA consulted with a range of nationally recognized climate scientists for their insights and recommendations regarding the most credible regional projections for use by State DOTs and local transportation agencies. As a result of these conversations, FHWA gained access to key data sets not previously published in their entirety, including data compiled from the CMIP3 database that underlies the USGCRP's *Global Climate Change Impacts in the United States* (2009) report. The tables of regional climate changes provided in this report were derived from an analysis of that data set. That information is complemented with the results of an extensive literature search and review, which is summarized in Appendix C. The search was conducted using a variety of databases of journal articles, government reports, and other peer-reviewed publications encompassing a variety of spatial scale information from regional- to city-scale.

The conclusions and data sets revealed through these efforts were further evaluated and scrutinized, and subsequently vetted with regional climate experts. A methodology was developed that identified which studies in the Climate Change Effects Typology Matrix correlated with optimum model characteristics, and these studies were then included in the regional narrative in the main report.

It should be noted that each study cited in this report has a unique set of model characteristics and associated uncertainty,[6] which makes comparing results across studies challenging. Although all of the differences in assumptions and approaches among these

---

[6] As described by Hawkins and Sutton (2009), model uncertainty, natural variability uncertainty, and scenario (i.e., GHG emissions scenario) uncertainty contribute to the total uncertainty associated with each projection. The structure of a modeling study helps define the associated uncertainty. For example, the magnitude of the uncertainty related to each of the three factors varies according to the time horizon of the projections. Hawkins and Sutton (2009) find that natural variability represents a large portion of the total uncertainty in applying climate projections in the *near-term*, dropping off significantly by *mid-century*. Model uncertainty is also a significant contributor to total *near-term* uncertainty and tends to stay relatively similar in magnitude through the projected century. For *mid-century*, the scenario and model uncertainties are somewhat similar in magnitude. For *end-of-century*, the scenario uncertainty contributes the greatest degree of uncertainty to the total. These uncertainties also change relative to each other when projections are provided at a smaller spatial scale (i.e., global to regional). For finer scale analysis, natural variability, in particular, significantly affects total uncertainty of a projection across all future time periods.

studies are not explicitly described in this report, care was taken to ensure that only logically comparable aspects of the studies are presented.

FHWA's methodology for identifying, categorizing, reviewing, and presenting projections of climate change effects in this report involved the following steps:

(1) **Selection of relevant climate effects**: The climate change effects determined to most affect highways and highway networks and discussed in this report include changes in average and extreme temperature, changes in average and extreme precipitation, and sea-level rise. Several recent reports have highlighted the importance of these effects with respect to the U.S. transportation system and highways in particular (NRC 2008; USGCRP 2009; CSIRO 2006; CIG 2007). Extreme precipitation is associated with storm activity such as convective storms, extratropical storms, or tropical storms. There is limited information available pertaining to storm activity projections that describes changes in storm intensity, frequency, and duration. Additional variables such as relative humidity, solar radiation, and extreme cold events are also relevant, but regional information for these variables was not available for inclusion in this report.

(2) **Literature review:** The compilation of regional projections began with a literature search using relevant, carefully composed search terms across relevant databases of publications on the environment, energy, technology, and government.

(3) **Screening literature review findings:** The search was refined to include articles, government reports, and peer-reviewed publications with a published date post-2003 and available by June 2009. Effort was made to include seminal reports that became available after this date. This approach was taken to ensure that the versions of the climate models used were less likely to draw from out-of-date parameterizations, and that the emission scenario projections used were more likely to be based on the IPCC SRES scenarios used in the Fourth Assessment Report.

(4) **Populating the Climate Change Effects Typology Matrix:** The literature was organized into the Climate Change Effects Typology Matrix (located in Appendix C) by U.S. region based on a recent panel-reviewed report (USGCRP 2009); by time horizon where *near-term* represents 2010-2040, *mid-century* represents 2040-2070, and *end-of-century* represents 2070-2100; by climate effects (listed in the order of temperature, precipitation, storm activity, and sea-level rise); and by spatial coverage.

(5) **Consultation with federal climate experts**: A group of national federal climate experts from organizations including NOAA's National Climatic Data Center, USGS, DOE and others provided guidance on the criteria used in this report for determining whether a study would be included in the Section 3 regional narratives. There was a strong consensus for providing a plausible range of projections (tied to the low B1 and moderately high A2 IPCC greenhouse gas emissions scenarios) as opposed to a single point value for the *mid-century* and

*end-of-century* projections. Projections of seasonal and annual temperature and seasonal precipitation compiled from CMIP3 database for use in the 2009 USGCRP report were used together with the results of the literature review (see Appendix A for more information on the emissions scenarios).

(6) **Additional data analysis and refinements to regional climate change data**: This report includes new analyses of the CMIP3 database of climate model integrations compiled by Michael Wehner of the Lawrence Berkeley National Laboratory and used in the USGCRP (2009) report.[7] Regional values are determined by using the corresponding grid cells of each climate model that fall within each region (Figure 6, for example, was developed using this collection of projected data). Then for each region and each of the three time frames, the following statistics have been computed for temperature and precipitation: "mean," "likely," and "very likely" (see below for a definition of terms). The results are included in regional tables in Chapter 3, the regional maps in Appendix B, and the Climate Change Effects Typology Matrix (Appendix C). These results provide mean conditions (as opposed to variability) at the regional scale.

(7) **Inclusion of downscaled data:** High-resolution temperature and precipitation projections for the continental United States developed through statistical downscaling of the results of 16 climate models of the CMIP3 database were provided to FHWA (Liang et al. 1994; Maurer et al. 2002).[8] The projections are provided for a low (B1) and moderately high (A2) emission scenarios for three future projections including *near-term* (2010-2039), *mid-century* (2040-2069), and *end-of-century* (2070-2099) relative to a 1971-2000 baseline. Figures of the temperature projections are provided in Appendix B, while figures of thresholds such as extreme temperature are provided within this report (Maurer et al. 2002; Maurer et al. 2008). Downscaled data may be preferred when projections are required for an area finer than the spatial resolution provided by the climate models; particularly if the location is not well-represented by the larger-scale averages. In addition, downscaling data provides a mechanism for translating larger-temporally scaled climate model projections to finer-temporally scaled climate variability.

(8) **Consultation with regional climate experts**: The Climate Change Effects Typology Matrix was vetted with regional climate experts to discuss studies included and identify any missing studies. Based on this review, some studies were removed from the matrix or placed into the national section. The consultation also included discussions of particular regional problematic climate effects. There was a general consensus that it is important to not "cherry pick"

---

[7] These values are averaged for each region from the corresponding grid cells of each climate model in the CMIP3 database. This is appropriate as the information provides mean conditions of temperature and precipitation at the regional scale. Statistical downscaling of the CMIP3 database for the continental United States is used in this report for the figures of extremes (such as the days the maximum temperature reaches or surpasses 90°F); the downscaling results are not available for locations outside the continental United States. See Appendix A for more information.

[8] Bias-corrected and spatially downscaled climate projections derived from CMIP3 data, described by Maurer et al (2007). We acknowledge the Program for Climate Model Diagnosis and Intercomparison (PCMDI) and the WCRP's Working Group on Coupled Modelling (WGCM) for their roles in making available the WCRP CMIP3 multi-model dataset. Support of this dataset is provided by the Office of Science, U.S. Department of Energy.

climate models for calculating the mean and ranges from the USGCRP data. Alaska is the exception where the general consensus was to draw results from the five top performers identified by the Walsh et al. (2008) study.

(9) **Treatment of uncertainty**: There is always some degree of uncertainty associated with model projections. The Climate Change Effects Typology Matrix includes this information, when included in the source study, in a column labeled "certainty." In general, the model projections are more uncertain the further they are into the future. A small range of uncertainty tends to exist in the *near-term* time horizon, while a larger range of plausible values exists for the *long-term*. The temperature and precipitation information from the USGCRP data set that are presented for each region were quantitatively analyzed in this study to characterize plausible future climate conditions and the associated uncertainty. Each scenario/model combination produces a single data point. There are 15 models run for the A2 emission scenario to 19 models run for the B1 emission scenario, producing 15 to 19 mean results for each variable in each time frame. The following information is provided for precipitation and temperature for each region (see Figure 2):

- "**Mean**" – The mean range is the average of all of the simulations in the lower emission scenario (B1) and the average of all of the simulations in the higher emission scenario (A2). It is a simple measure of the central tendency of the projections and the uncertainty associated with future greenhouse gas (GHG) emission rates.

- "**Likely**" – The likely range is computed by first determining the standard deviation above and below the mean for each scenario.[9] Then, the minimum and maximum of these four values (i.e., two from each scenario) are defined as the likely range. The range is a measure of the differences (and uncertainty) associated with the models that were used, as well as the uncertainty of future GHG emission rates.

- "**Very Likely**" – The very likely range is computed in the same way as the likely range, except that two standard deviations are used instead of one.[10]

---

[9] Assuming the data are well represented by a Gaussian distribution, the likely range represents about 68% of the values extending from the 15$^{th}$ percentile to the 85$^{th}$ percentile.

[10] Assuming the data are well represented by a Gaussian distribution, the very likely range represents about 95% of the values extending from the 2.2$^{th}$ percentile to the 97.8$^{th}$ percentile.

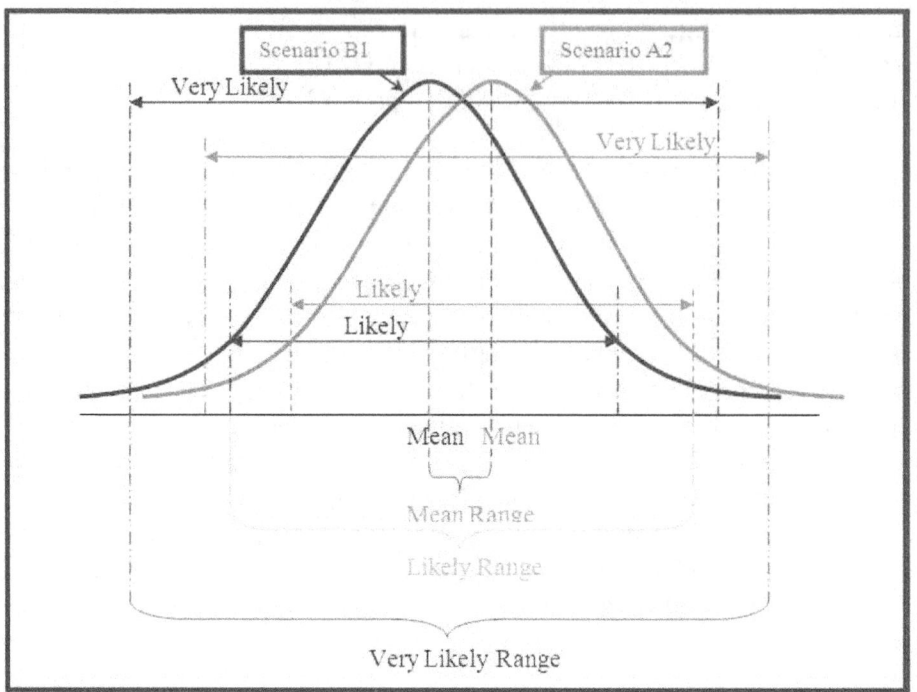

Figure 2: This figure describes the process of creating the mean, likely, and very likely ranges from the USGCRP data and reported in Section 4. A curve represents an idealized version of connecting the points of results obtained from a variety of climate models. The blue curve and associated labels represent the mean, likely, and very likely values for a given emission scenario; likewise for the green curve assuming a Gaussian distribution. Each curve describes model uncertainty for the given emission scenario. The ranges provide the uncertainty associated with both the emission scenario and the climate models.

(10) **Developing Regional Narratives for Section 3:** A methodology was developed to synthesize the array of projections provided by the Climate Change Effects Typology Matrix into a regional narrative discussion that can assist in informing future analysis of climate impacts on the highway system. The criteria used in this report for determining whether a study would be included in the regional narratives of Section 3 were guided by discussions with climate experts and are illustrated in Figure 3.

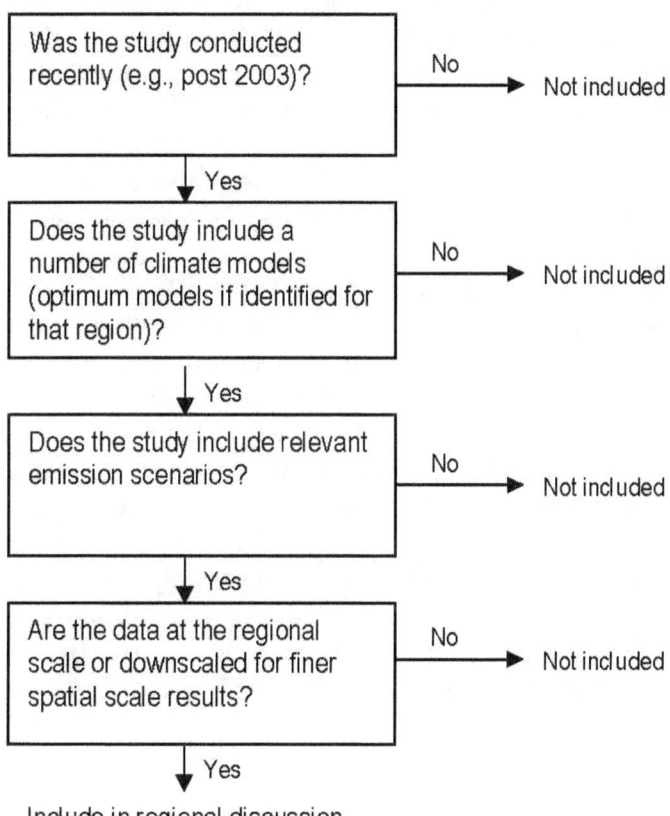

**Figure 3: Criteria used to assess whether a study is included in the narrative regional discussions.**

# 3 Projected Climate Change by Geographic Region

This section begins with a brief review of national-scale changes in temperature, precipitation, sea level, and storm activity. This information is drawn largely from the USGCRP (2009) report. These national-scale descriptions are followed by the core of this report: a series of sections providing the current state of the knowledge concerning the aforementioned variables for every region in the country.[11] When possible, related regional-scale information about extreme climate stressors is also provided to assist highway planners in relating climate change impacts on infrastructure of concern. For example, a given projection of changes in extreme events such as increases in heavy precipitation or, conversely, drought, may stress a given infrastructure if the projected conditions fall outside the design range. (The Climate Change Effects Typology Matrix in Appendix C provides detailed information of the entire, and broader, set of studies collected for this effort.) After discussion with climate experts, a set of inclusion criteria was produced that determined which of the collected studies would be discussed in this section (Section 2 and Appendix A provide further discussion of this step). As discussed in Section 2, there is a large degree of uncertainty associated with climate projections and, accordingly the projections are compiled from a number of climate models and across emission scenarios to help bracket the range of plausible futures. These projections are provided at varying degrees of uncertainty: "mean range," the "likely range," and the "very likely" range (as discussed in Section 2). Temperature projections for all ranges and regions are projected to rise. The plausible ranges for precipitation projections, however, may extend from a reduction to an increase, which complicates the explanation of regional precipitation patterns.

## 3.1 National projections

The following bullets provide a general overview of how temperature and precipitation are projected to change for the United States (USGCRP 2009):

- Temperatures will continue to warm over the century with a projected average increase by the end of the century of approximately 7 to 11° F under the high emission scenario and 4 to 6.5°F under the low emission scenario.[12]

- For the contiguous United States, summer months are projected to experience greater warming nationally compared with winter months. Extreme heat days (defined as a daily maximum temperature that currently has a 5% chance of occurring each year) will grow in number while extreme cold will decrease. By the end of the century, extreme heat events are projected to have a 50% chance of occurring each year.

---

[11] It should be noted that the time ranges used in the tables and maps (and the USGCRP data) are slightly different than those defined for this report. While the USGCRP data are representative of the data included under the time periods as defined in this report, the time periods are not an exact match.

[12] These ranges correspond to the mean averages as defined in this report, demonstrating differences between climate model results based on a given emission scenario.

- Heavy downpours are projected to continue to increase while the lightest precipitation decreases. By the end of the century, heavy downpours that have a 5% chance of occurring in a given year are projected to have a 20 to 75% chance of occurring in a given year. In addition, these types of precipitation events are projected to be 10 to 25% heavier.

- In general, northern areas of the country are projected to become wetter while southern areas, particularly the Southwest, will become drier.

- The jet stream over the United States is expected to continue to shift northward. The jet stream steers and fuels mid-latitude storms (i.e., extratropical storms).

- Precipitation and humidity are projected to significantly increase in the northern United States during winter and spring. This increase is in response to warm, moist air from the south moving northward and replacing very cold, dry air masses.

- Warming temperatures will increase evaporation as well as shift the rain/snow line northward and to higher elevations so more precipitation will fall as rain and less as snow.

### 3.1.1 Temperature

The likely range for global annual mean temperatures is projected to rise this century by 2 to 11.5°F by 2100; this range is based on the multi-model ensemble results across each of the six SRES (USGCRP 2009; IPCC 2007a).[13] This warming is not projected to be evenly experienced around the globe. The greatest warming is projected to occur over land and in most high northern latitudes (IPCC 2007b). A number of studies suggest that irreversible, severe, and widespread impacts would be associated with a 2°F increase in average global temperatures above 1980-1999 levels (USGCRP 2009).

Within the United States, the annual mean temperatures by the end of the century are projected to warm by approximately 7 to 11°F under the higher A2 emissions scenario and by approximately 4 to 6.5°F under the lower B1 emissions scenario (USGCRP 2009). Summer months are projected to experience greater warming nationally compared with winter months. Figure 4 illustrates the warming projected for *mid-century* and *end-of-century* under a higher (A2) and lower (B1) emissions scenarios for the continental United States. In addition to increased mean summer temperatures, extreme heat days will grow in number while the number of extreme cold days will decrease. By the end of the century, extreme heat events that currently have a 5% chance of occurring each year are projected to have a 50% chance of occurring each year under a moderate (A1B) emission scenario (USGCRP 2009). This study further finds that in addition to more

---

[13] The IPCC definition for "likely range" is -40% to +60% around the mean for each SRES. The full likely range provided in this report is from the lowest point and highest point of all the likely ranges.

frequent occurrence of extreme heat waves, very hot days are projected to be about 10°F hotter than they are today (USGCRP 2009).

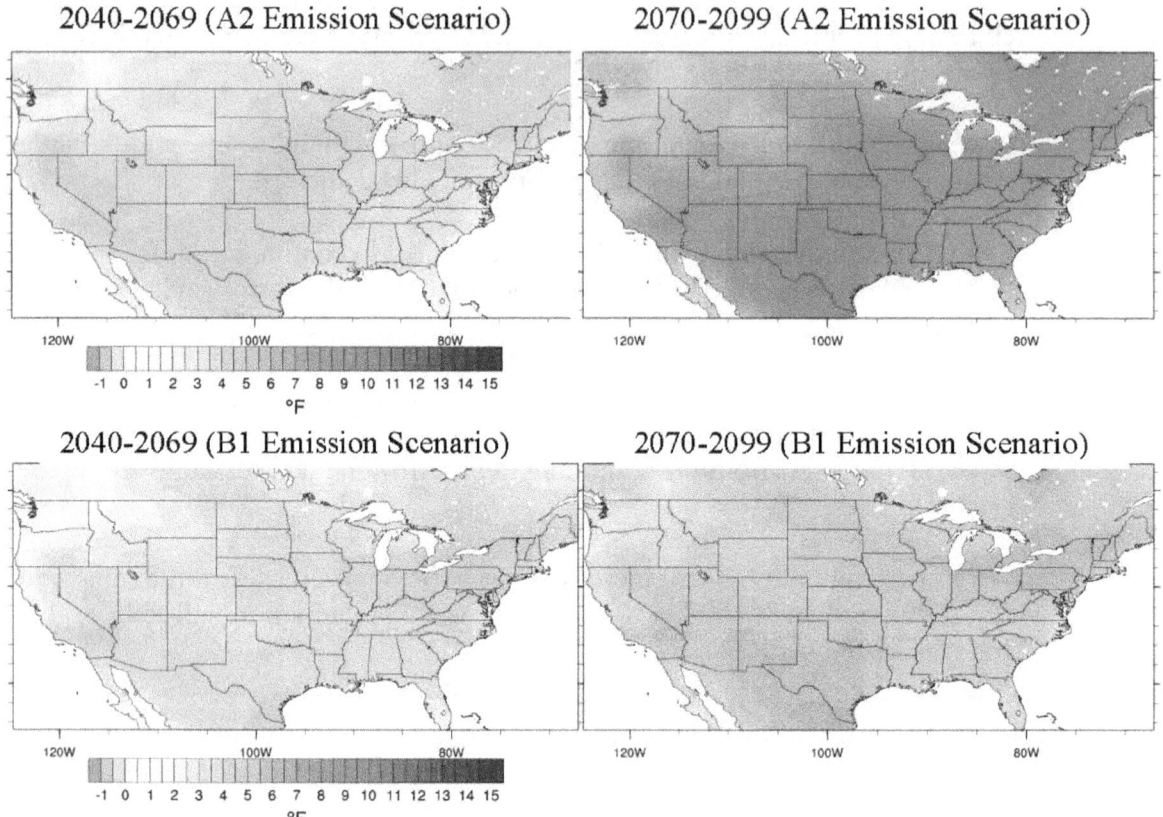

Figure 4. Projected mean summer temperature change (°F) relative to 1971 to 2000 based on the projections of statistically downscaled CMIP3 projections (see Appendix B for additional temperature projections).[14]

### 3.1.2 Precipitation and Storm Events

Similar to temperature, projected precipitation is discussed in terms of seasonal averages and extreme precipitation events (i.e., heavy downpours). Discussion of annual precipitation, however, is limited in this report as it masks important variability, and provides, in some cases, misleading information. Seasonal precipitation, on the other hand, illustrates important trends while smoothing out the effects of heavy downpours. At a daily scale, discussions of heavy precipitation events, generally defined as greater than 2 inches per day, provide information about isolated storm activity;[15] however, no processed information is readily available regarding other storm variables such as wind strength and direction, nor the type of storm causing the event. The combination of

---

[14] These figures were provided by personal communication with Katharine Hayhoe, Texas Tech University and produced for USGCRP (2009).
[15] It should be noted that 2 inches per day of precipitation may not be considered a threat to highway infrastructure; however, a few studies cited in this report provide projections for the increased frequency of the 95th percentile of precipitation.

projected seasonal precipitation and precipitation events provide compelling information for system planning.

There is greater uncertainty associated with future changes in total precipitation compared with temperature, because precipitation is more heavily influenced by both small-scale phenomena and climate variability not captured by climate models. The uncertainty represented in sections 3.3 through 3.11 through the provided *likely* and *very likely* ranges encompasses the extent of plausible seasonal precipitation futures. The USGCRP report (2009) states that the confidence of seasonal precipitation projections for the United States is highest for winter and spring when precipitation is projected to increase significantly for the northern region as the boundary between the southern warm moist air and the northern cold air shifts northward. In addition, the northern regions are projected to experience more precipitation falling as rain and less as snow. Conversely, the southern regions are projected to experience significant reductions in precipitation during the winter and spring months, particularly in the Southwest.

Extreme precipitation events are projected to increase in frequency and intensity, while the amount that falls in light precipitation events is projected to decrease (USGCRP 2009). By the end of the century, heavy downpours that have a 5% chance of occurring in a given year are projected to have a 7 to 25% chance of occurring in a given year, depending on location (USGCRP 2009). In addition, heavy downpours that have a 5% chance of occurring today are projected to be 10% heavier under the lower emission scenario (B1) to 25% heavier under the higher emission scenario (A2) than it is now (USGCRP 2009). An apparent paradox of increased moisture leading to both increased drying conditions and increased heavy precipitation events is actually consistent with a warmer atmosphere. As temperatures increase, the air can hold more water vapor, allowing for increased amounts of evaporation. As more moisture enters the atmosphere, rises, and cools aloft, the water vapor condenses back to a liquid, leading to a greater amount of precipitation and increasing the energy associated with the storm (i.e., energy is released when water vapor condenses to a liquid).

The United States is home to an impressive and diversified set of storms. In the Great Plains, for example, convective storms can become so severe as to produce damaging strong winds, large-sized hail, and tornadoes. In the Southeast, convective storms in Florida produce a significantly high number of lightning strikes per year. In the Northeast, particularly New England, nor'easters[16] can produce intense and damaging conditions. The Gulf states and the Southeast are vulnerable to tropical storms and hurricanes. The Pacific coast experiences coastal storm and flooding events resulting from the (in)famous Pineapple Express.[17]

The storm events currently experienced in the United States will likely evolve in response to a changing climate. Though the current research is somewhat insufficient to draw

---

[16] A type of extratropical storm that often initially develops as a cold-core low pressure system near the Gulf of Mexico, gathering warmth and moisture, then travels northward, developing along the East Coast.
[17] Pineapple Express occurs when humid subtropical air originating near Hawaii travels to California causing great rains and floods.

conclusive projections, some broad relationships between the plausible future conditions and the impacts these conditions may have on storm development have been discussed in the CCSP's (2008b) *Weather and Climate Extremes in a Changing Climate* report. This discussion provides information for three types of storm events: convective storms (e.g., thunderstorms), extratropical storms (e.g., cyclonic storms forming along a mid-latitude or high-latitude front), and tropical storms and hurricanes[18] (e.g., organized thunderstorms with cyclonic motion originating in the tropics).

Convective storms that are very localized may in fact increase in intensity in response to the increase in atmospheric moisture, but decrease in duration or frequency. Projections of changes in convective storms are unclear and may improve with the application of nested models (regional models driven with GCM data that in turn feed information back to the GCM model).

The physical mechanisms associated with extratropical storms are not yet entirely understood even for present-day events; for example, major El Niños are understood to influence storm behavior, but it is unclear how the natural variability of other large-scale circulations affects these storms. To complicate matters, the characteristics such as sea-level pressure or strong surface winds used to define an extratropical storm differ between climate studies. Even under these circumstances, some consistent changes across studies of extratropical storms have been identified and suggest strong storms will be more frequent, while the overall number of storms may decrease.

The recent scientific consensus on tropical cyclonic activity (i.e., tropical storms and hurricanes, also known as typhoons in the Pacific, a cyclone in India, and a tropical cyclone in Australia) describes the globally averaged intensity of tropical cyclones as increasing by 2 to 11% by the end of the century (Knutson et al. 2010). On the other hand, the globally averaged frequency of tropical cyclones is consistently projected by modeling studies to decrease by 6 to 34%; however, results from higher resolution models suggest increases in both the frequency of intense storm activity and the precipitation rate within the storm center. It remains uncertain whether past changes in tropical storm activity are influenced by natural variability or human activity (Knutson et al. 2010; CCSP 2008b). It should be noted that precipitation presented by region in the following subsections encompasses changes in precipitation associated with storm events.

### 3.1.3 Sea-Level Rise

Detailed national information is not uniformly available for sea-level rise projections. Many of the state-of-the-science studies discussing sea-level rise provide only global projections. A number of factors determine sea level changes at any given location. On the global scale, the two most important factors are the expansion of ocean water as it warms, and changes in the amount of water in the oceans due to the accumulation and melting of ice sheets and glaciers and changes in the amount of water stored in

---

[18] A tropical storm, by definition, becomes a hurricane when sustained winds reach 74 miles per hour. A tropical depression, tropical storm, and a hurricane are all types of tropical cyclones.

reservoirs.[19] These factors are generally considered in the production of global sea level projections.

According to the Intergovernmental Panel on Climate Change (IPCC) Fourth Assessment Report (AR4), globally averaged sea level will rise by at least 5" by *mid-century* and 7" to 23" by *end-of-century* (IPCC 2007a). Due to significant uncertainty associated with future changes (e.g., melting rates) in the volume of glaciers and ice sheets at the time that report was written, the IPCC essentially excluded major contributions from those factors in its quantitative projections of sea level. Methods have been developed since the publication of the IPCC AR4 results that attempt to address these issues, though comparisons across recent studies can be difficult due to differing analytical approaches and Earth system components that are included in each study. Table 3-1 describes the global sea-level rise projections estimated by a number of recent studies for the end of the century, providing a range of projected global sea-level rise of 7 to 79". It is obvious that there is great variability in the results, but note that most of the estimates are significantly higher than those by the IPCC (2007a). The low end of the IPCC (2007a) projections is considered very conservative, for example, and assumes negligible contribution from melting of Greenland (DWR 2008). The continued rise in global sea level would greatly affect coastal locations, particularly those already vulnerable to storm surge. The impact of sea-level rise could be further exacerbated if coastal landforms that serve as storm-surge barriers are lost to extreme storm events such as a hurricane (USGCRP 2009).

Locally, other factors including vertical land motion (i.e., subsidence or uplift of land), sedimentation and erosion, ocean circulation, gravitationally induced changes, and ocean density (affected by regional changes in ocean salinity and ocean temperature) can also play a role (with vertical land motion often dominant), thus complicating the work of making projections at the sub-global level. Many of these factors are not well understood and are the subject of current research efforts, so any regional or local projections made at this time are fraught with uncertainty and should be considered carefully. It should be noted that the term "relative sea-level rise" refers to the changes in land elevation with respect to the level of the ocean determined by tide gauge measurements. Several different modeling and analytical tools are used to predict changes in each of these factors, which must be summed to determine local changes. Estimates of regional and local relative sea-level rise have been collected and are discussed in sections 3.3 through 3.11, as available. It should be noted there is no study that considers all these factors, nor is there a consistent methodology applied across these studies, so local projections should be considered carefully.

---

[19] "Steric" sea-level change refers to changes in sea level due to thermal expansion and salinity. "Eustatic" sea-level rise refers to the changes in sea level in response to the melting of small glaciers and ice sheets. "Isostatic" sea-level change is due to the shifting of land masses through adjustment to glacial loading or unloading, thermal buoyancy, or plate tectonics. "Dynamic" sea-level change is due to changes in ocean circulation.

| Study | Projection, 2100 | Methodology |
|---|---|---|
| IPCC (2007a) | 7" to 23" (18cm to 59cm) | Accounts for thermal expansion and conservative estimates of changes in ice/snow melt. |
| Rahmstorf (2007) | 20" to 55" (0.5m to 1.4m) | Assumes a linear relationship between 20$^{th}$ century observed temperature and sea-level rise to obtain a proportionality constant of 3.4 mm/year per °C that was used in Rahmstorf's estimate of future sea level. This projection relies upon the assumption that the past statistical relationship remains constant in the future and uses the global mean temperature projections of the IPCC Third Assessment Report (TAR) were used. |
| Grinsted et al. (2009) | 35" to 51" (0.9m to 1.3m) | Uses four inversion experiments to relate 2,000 years of global temperatures to sea level and validated model parameters with satellite altimetry. The global mean temperature projections of 6 IPCC AR4 emission scenarios were used. |
| Rohling et al. (2008) | 63" (1.6m) | Uses paleoclimate data of the last interglacial period, when global mean temperatures were at least 2°C warmer than today and comparable to current projected temperatures. |
| Pfeffer et al. (2008) | 31" to 79" (0.785m to 2.008m) | Uses thermal expansion projected by IPCC AR4 together with kinematic scenarios (e.g., varying the velocities of outlet glaciers) to estimate the change in surface mass balance of ice of Greenland and Antarctica, and discharge of melting ice sheets and glaciers. |

Table 3-1: Global projections of sea-level rise compared with 1990 levels, except Rohling et al. (2008) which describes projections per century. All results have been converted to inches, with study results provided in parenthesis. These studies do not account for additional regional factors that may cause regional sea-level change to be greater or less than the global average.

The IPCC factored in regional changes of ocean density and ocean circulation to estimate local sea-level change. Figure 5 illustrates the variation in sea-level rise when these factors are considered. Local sea-level changes will also be affected by other changes in smaller scale circulation patterns, and vertical land movement, not included in this figure.

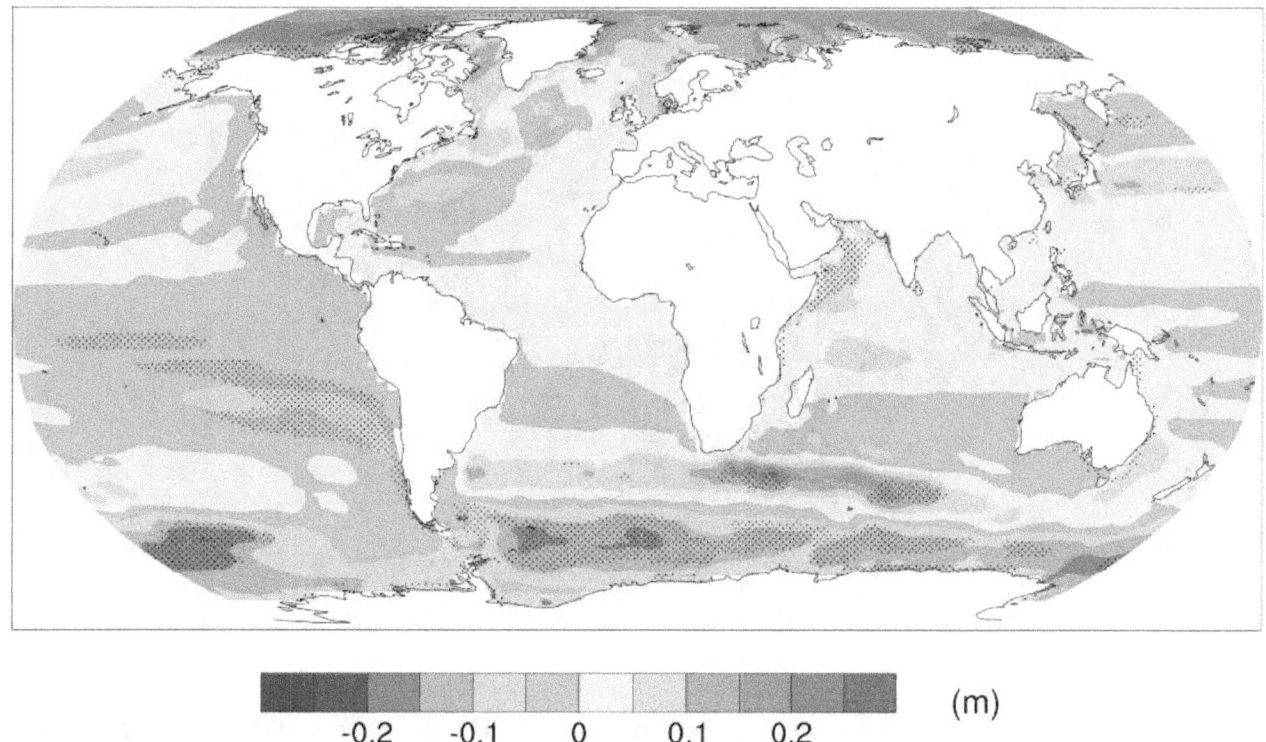

Figure 5: The intent of this figure is to simply illustrate the potential variability of regional differences, and should not be relied on for planning purposes. This figure shows projected local sea-level change (in meters) due to ocean density and circulation change relative to the global average (i.e., positive values indicate greater local sea-level change than the global average) during the 21st century, calculated as the difference between averages for 2080 to 2099 and 1980 to 1999, as an ensemble mean over 16 AOGCMs forced with the SRES A1B scenario. Stippling denotes regions where the magnitude of the multi-model ensemble mean divided by the multi-model standard deviation exceeds 1.0 (IPCC 2007a). This figure does not include other local factors such as land uplift or subsidence.

### 3.1.4  Local Applications of Regional Data

Most of the regional projections provided in sections 3.3 through 3.11 are averaged across each region. Due to regional terrain or other phenomena, local variability within a region may be large and may affect the robustness of using an averaged regional projection. For example, Figure 6 demonstrates the large local variability in extreme heat events[20] projected to occur at the end of the century for the contiguous United States.[21] In this example, a regional average of the inter-mountain West would only be a very rough approximation of the local conditions.

---

[20] In this case, an extreme heat event is defined as a day where the maximum temperature exceeds 90°F.
[21] Statistically downscaled data from the CMIP3 database are used for this figure.

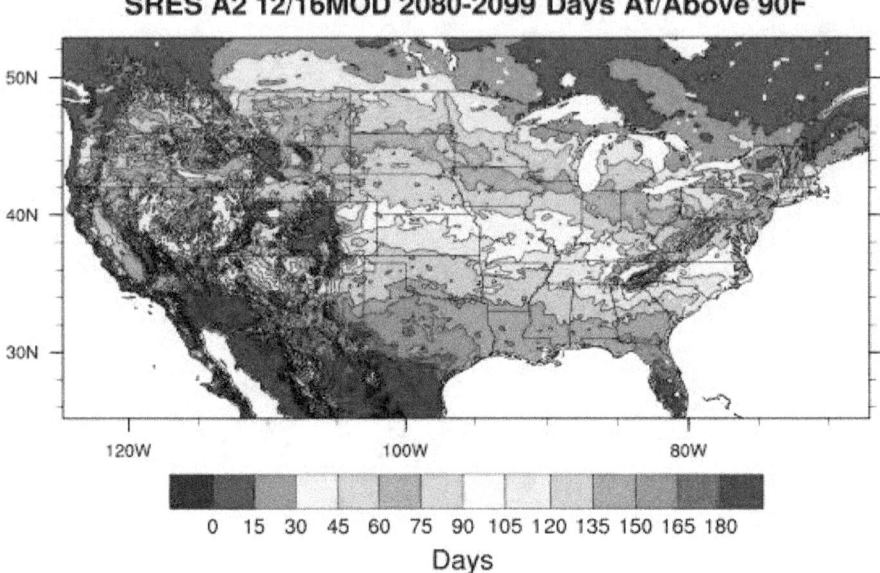

Figure 6: Extreme heat days (days where the maximum temperature exceeds 90°F) (USGCRP 2009).

The confidence of these projections also varies regionally. Figure 7 presents projected changes in annual average precipitation for North America by 2080-2099 relative to precipitation in the recent past (USGCRP 2009). The hatched areas on the maps demonstrate projections where confidence is highest (that is, at least two out of three models agree on the sign of the projected change in precipitation). According to Figure 7, during the summer months, two large pockets of considerable reduction in precipitation are evident within the continental United States, though there is only high confidence in the Pacific Northwest drying. Overall, confidence is higher for the winter and spring seasons where the northern regions are projected to experience significantly more precipitation in response to the northward movement of the boundary between warm, moist southern air and cold, continental northern air (USGCRP 2009).

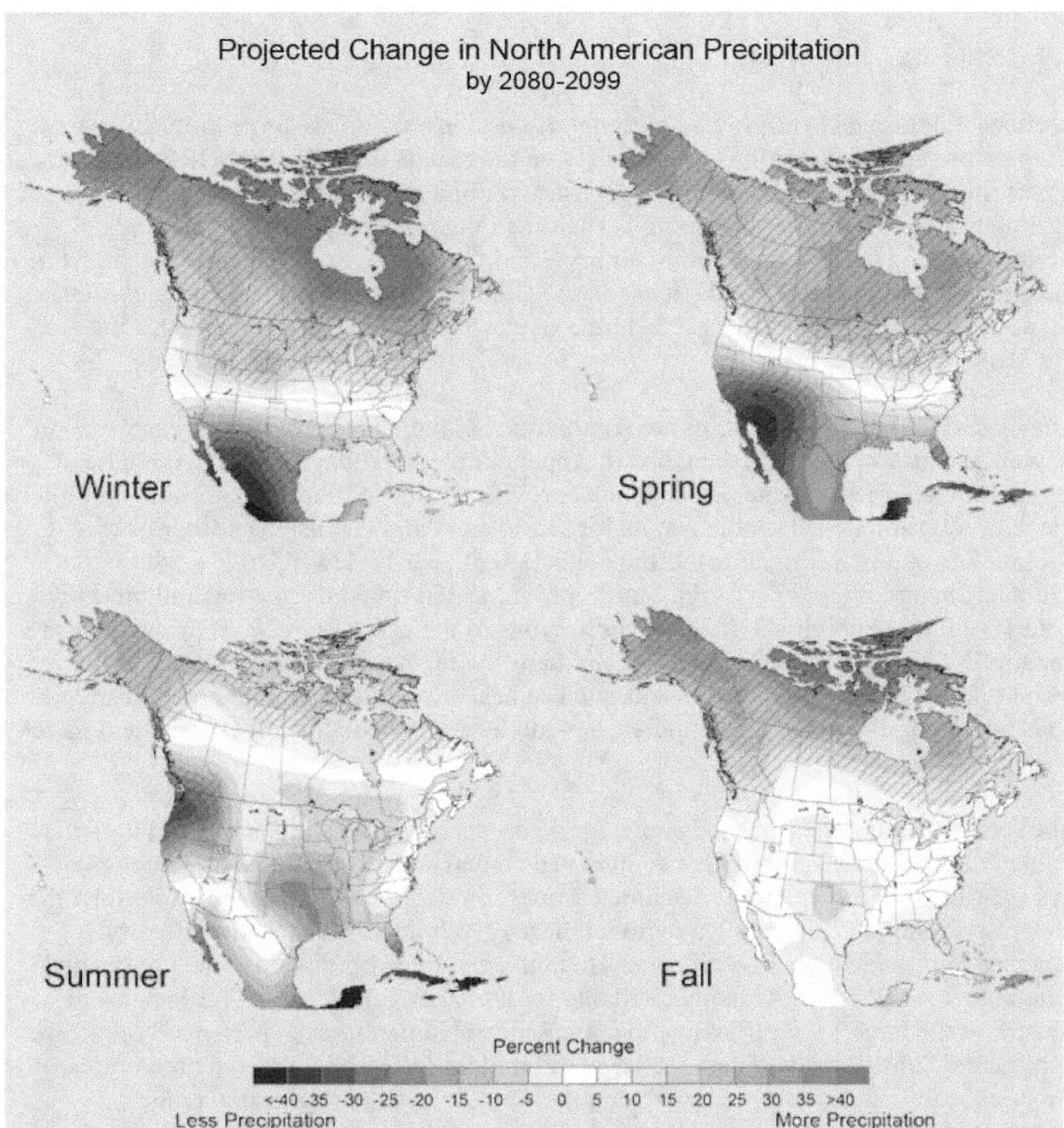

Figure 7: Projected future changes in precipitation by 2080-2099, relative to average seasonal precipitation 1961-1979 under the A2 emission scenario and simulated by 15 climate models (CMIP3 data; USGCRP 2009). Hatched areas show areas with highest confidence in the projected change.

## 3.2 Regional Summaries

Sections 3.3 through 3.11 provide regional discussions of climate projections of temperature and precipitation based largely on the results from the USGCRP (2009) report (however, the USGCRP report focuses on impacts and does not provide this information uniformly for all regions). The time ranges presented in the regional summaries (i.e., "near-term", "mid-century", and "end-of-century") are developed from the collection of peer-reviewed studies in the Climate Change Effects Typology Matrix. As noted earlier, the USGCRP (2009) time ranges fall within these time ranges but are not identical to them.

The USGCRP (2009) projections are summarized in the tables provided in each section as well as illustratively with the maps in Appendix C (these maps also illustrate the regional boundaries). The tables and maps provide the mean, likely range, and very likely range of multi-model ensemble results for "low" and "high" emission scenarios (see Section 2 for more information). If the collection of peer-reviewed studies within the Climate Change Effects Typology Matrix provide additional information and meet the criteria outlined in Figure 3, then the discussions of the USGCRP (2009) projections are enhanced with these additional studies. In many cases, these additional studies provide insight into changes of extreme events such as heat waves. It should be noted this report does not attempt to judge these studies, beyond meeting the inclusion criteria, and merely presents them.

The regional summaries provide climate projections of temperature and precipitation. Table 3-2 and Table 3-3 provide the observed annual and seasonal means averaged over the 1961 to 1979 time period.[22] The means are provided as whole numbers to reflect the lack of precision in calculating a regional mean, given insufficient station density required to provide higher confidence. This information is provided for the continental United States and was not readily available for the other U.S. regions. Table 3-2 illustrates the large 15°F difference in annual mean temperatures experienced across the continental United States. There are large differences in seasonal mean temperatures experienced within each region, with summers exhibiting the warmest average temperatures and winters the coldest, with fall being slightly warmer than spring months.[23] The change in the observed annual mean temperature averaged over 1993 to 2008 relative to the 1961-1979 baseline is also provided in the continental regional summaries (likewise for the observed annual mean precipitation).

---

[22] This information was provided by personal communication with Jay H. Lawrimore of the National Climatic Data Center.
[23] Seasons are defined as follows: Winter (December, January, February), Spring (March, April, May), Summer (June, July, August), and Fall (September, October, November).

| Region | Mean Temperature (°F) | | | | |
|---|---|---|---|---|---|
| | Annual | Winter | Spring | Summer | Fall |
| Southeast | 63 | 47 | 63 | 78 | 65 |
| Northeast | 47 | 24 | 45 | 67 | 50 |
| Midwest | 47 | 21 | 46 | 69 | 50 |
| Great Plains | 52 | 29 | 51 | 73 | 53 |
| Southwest | 55 | 39 | 53 | 72 | 57 |
| Pacific Northwest | 47 | 31 | 45 | 63 | 48 |

Table 3-2. 1961-1979 annual and seasonal mean temperature (°F) for the continental U.S. regions.

Table 3-3 provides observed mean precipitation for each continental U.S. region. Annually, the Southeast experiences the greatest amounts of precipitation. The Southwest is particularly drier in comparison. Overall, winter tends to be a drier season for most regions, with the other seasons somewhat comparable to each other. The Pacific Northwest is an exception with winter as the wettest season and summer as the driest.

| Region | Mean Precipitation (inches) | | | | |
|---|---|---|---|---|---|
| | Annual | Winter | Spring | Summer | Fall |
| Southeast | 50 | 11 | 12 | 15 | 11 |
| Northeast | 41 | 9 | 10 | 11 | 10 |
| Midwest | 34 | 5 | 10 | 11 | 8 |
| Great Plains | 21 | 2 | 6 | 7 | 5 |
| Southwest | 15 | 5 | 3 | 3 | 4 |
| Pacific Northwest | 28 | 11 | 6 | 3 | 7 |

Table 3-3. 1961-1979 annual and seasonal precipitation mean (inches) for the continental U.S. regions. The total seasonal amounts may not equal the annual amount provided due to rounding.

Sections 3.3 through 3.11 also provide regional summaries of studies investigating local sea-level rise. In addition, a comparison of historical trends of the region[24] to the global observed trend of sea-level rise of $1.8 \pm 0.5$ mm yr$^{-1}$ from 1961 to 2003 (IPCC 2007a) is provided to illustrate past regional difference from the global average sea-level rise.

---

[24] The National Water Level Observation provides historic sea level trends for 128 stations along the U.S. coastline. These measurements are provided by NOAA and include stations that provide sea level trends over a 30 year span or longer (NOAA 2010).

## 3.3 Northeast

### 3.3.1 Temperature

#### 3.3.1.1 Near-term (2010-2040)

Within the next several decades, the Northeast is likely to experience an increase in annual mean temperature of approximately 2.5°F with a likely range of 1.9 to 3.2°F (USGCRP 2009; NECIA 2006; Frumhoff et al. 2007). This projected warming is greater than the 1.5°F increase experienced over the 1993-2008 time period when compared to a 1961-1979 baseline (USGCRP 2009).[25] Winter temperatures over the same time period are projected to increase even more, by approximately 3.0°F with a likely range of 1.8 to 3.8°F. *Near-term* summer and spring temperature increases are projected to be slightly greater than 2.0°F (USGCRP 2009). These results are consistent with projections for Pennsylvania (NECIA 2008), while the NECIA (2006) report, which uses similar but not identical emission scenarios (the high emission scenario is A1Fi), projects similar mean temperature increases for annual and summer. However, NECIA (2006) projects that the winter mean temperature is projected to experience a slightly higher increase of 3.3 to 3.4°F with a likely range of 2.5 to 4.0°F.[26] These temperature increases could lead to further reduction in the thickness and duration of winter ice on lakes and rivers, more precipitation falling as rain rather than snow, and earlier spring snowmelt affecting the timing of peak river flows.

The number of extreme heat days is also projected to increase across a number of Northeast cities. Buffalo, NY is projected to experience the smallest increase (of 2 to 5 days per year) and Philadelphia, PA and Pittsburgh, PA are projected to experience the greatest increase (9 to 11 days per year) (NECIA 2006).[27] Boston, MA is projected to increase experience 4 to 8 additional days per year above 90°F (USGCRP 2009; Hayhoe et al 2008).[28] Other cities projected to experience 5 to 10 more days per year of extreme heat include Concord, NH; Manchester, NH; Hartford, CT; and New York City, NY (NECIA 2006). Areas of Pennsylvania could experience more than a doubling of the frequency of extreme heat days (NECIA 2008).

---

[25] This information was provided by personal communication with Jay H. Lawrimore of the National Climatic Data Center.
[26] The NECIA (2008) and NECIA (2006) study uses statistical downscaling of the results of three climate models: CM2.1, HadCM3, and PCM, relative to a 1961-1990 baseline.
[27] Extreme heat day is defined by this study as the number of days with temperature above 90°F.
[28] Hayhoe et al. (2008) provides results based on the A1Fi (high) and B1 (low) emission scenarios; the climate model results of CM2.1, HadCM3, and PCM are statistically downscaled.

### 3.3.1.2 Mid-century (2040 – 2070)

By *mid-century*, the increase in annual mean temperature for the Northeast is projected to be between 3.8 to 4.8°F with a likely range of 2.8 to 5.8°F (USGCRP 2009). This range is also representative of the increase projected for the summer and fall seasons (USGCRP 2009). The Northeast is projected to experience greater warming in the winter, with temperature increases projected of 4.0 to 5.4°F with a likely range of 2.9 to 6.6°F (USGCRP 2009). The temperature increase during the spring months is slightly lower than the annual average at 3.5 to 4.1°F, with a likely range of 2.2 to 5.5°F (USGCRP 2009). The frequency of extreme heat days in Northeast cities is also likely to rise, with Boston, for example, seeing an additional 12 to 29 days over 90°F (USGCRP 2009). Seven individual Northeast cities examined are projected to experience an increase of approximately 8 to 39 extreme heat days per year (NECIA 2006). The northern cities tend to be represented by the low end of this range and the southern cities by the high end.

| Northeast ($\Delta$ Temperature) | | Near-term (°F) | Mid-century (°F) | End-of-century (°F) |
|---|---|---|---|---|
| **Annual** | Mean | 2.5 | 3.8 – 4.8 | 5.4 – 9.0 |
| | Likely | 1.9 – 3.2 | 2.8 – 5.8 | 4.2 – 10.8 |
| | Very Likely | 1.3 – 3.8 | 1.9 – 6.8 | 3.0 – 12.5 |
| **Winter** | Mean | 2.8 – 3.0 | 4.0 – 5.4 | 5.9 – 9.3 |
| | Likely | 1.8 – 3.8 | 2.9 – 6.6 | 4.7 – 11.0 |
| | Very Likely | 0.9 – 4.7 | 1.8 – 7.9 | 3.5 – 12.8 |
| **Spring** | Mean | 2.0 – 2.2 | 3.5 – 4.1 | 5.0 – 8.1 |
| | Likely | 1.2 – 3.0 | 2.2 – 5.5 | 3.6 – 10.0 |
| | Very Likely | 0.4 - 3.8 | 0.9 – 6.8 | 2.3 – 11.9 |
| **Summer** | Mean | 2.3 – 2.5 | 3.7 – 4.8 | 5.2 – 9.4 |
| | Likely | 1.8 – 3.1 | 2.8 – 5.8 | 3.9 – 11.8 |
| | Very Likely | 1.3 – 3.7 | 1.8 – 6.9 | 2.7 – 14.1 |
| **Fall** | Mean | 2.5 – 2.7 | 3.9 – 4.8 | 5.3 – 9.1 |
| | Likely | 1.9 – 3.3 | 2.8 – 5.6 | 3.9 – 10.8 |
| | Very Likely | 1.2 – 3.9 | 1.8 – 6.5 | 2.5 – 12.8 |

Table 3-4: Annual and seasonal temperature changes for the Northeast region over the *near-term* (2010-2029), *mid-century* (2040-2059) and *end-of-century* (2080-2098) relative to 1961-1979. The range values are from low (B1) and high (A2) emissions scenarios. Data are from the USGCRP (2009).

### 3.3.1.3 End-of-century (2070-2100)

By the *end-of-century*, the warming in the Northeast is likely to be quite significant. The projected annual mean increase is expected to be between 5.4 to 9.0°F, with a likely range of 4.2 to 10.8°F. The winter, summer, and fall changes are all relatively similar to the annual mean (USGCRP 2009). The spring months are projected to experience the lowest mean increase of all the seasons, with a mean warming of 5.0 to 8.1°F and a likely range of 3.6 to 10.0°F (USGCRP 2009). Though the NECIA (2006) study estimates similar projections, the Pennsylvania study suggests greater increases, with winter temperature projected to rise 8°F and summer temperatures projected to rise 11°F. Not only are temperatures projected to change, but also the duration of each season. Overall, under a "business as usual" scenario (A2), winters in the Northeast are projected to shorten, with the length of the winter snow season cut in half for the northern half of the region, including New York, Vermont, New Hampshire, and Maine, and reduced to a few weeks in the southern half (USGCRP 2009). Summer-like temperatures, on the other hand, are projected to persist for 6 weeks longer than usual (USGCRP 2009).

Extreme heat events[29] that currently have a 5% chance of occurring each year are projected to have a 50% chance of occurring each year by late century (USGCRP 2009). Many Northeast cities are projected to experience approximately 13 to 63 more days reaching 90°F by the end of this century compared with today's observations (NECIA 2006). The northern cities in the region tend to fall in the low end of this range and the southern cities in the high end.

### 3.3.2 Precipitation and Storm Events

#### 3.3.2.1 *Near-term* (2010-2040)

Current observations averaged over the 1993 to 2008 time period suggest that annual mean precipitation has increased by 7% relative to the 1961-1979 time period (USGCRP 2009).[30] Within the next several decades, the Northeast is likely to experience wetter winter months with an average precipitation increase of about 6% and a likely range of +2 to +11% (USGCRP 2009). The fall months have the lowest mean projected increases of 1 to 2% with a likely range of -4 to +6% (USGCRP 2009). The spring and summer are projected to experience similar *near-term* increases in precipitation. Spring precipitation is projected to increase by 3% with a likely range of -2 to +7%, and the summer months projected to increase by 2% with a likely range of -1 to +6% (USGCRP 2009).

Understanding how precipitation intensity, duration, and frequency may change is important for planning purposes. Individual precipitation events are likely to increase in intensity by approximately 7% (NECIA 2006), so more rain may be arriving in brief pulses. Specifically, the maximum amount of precipitation to fall during any five-day period in a year is projected to increase by 9 to 12% (NECIA 2006).

---

[29] Extreme heat event is based on apparent temperature, a combination of high temperature and humidity.
[30] This information was provided by personal communication with Jay H. Lawrimore of the National Climatic Data Center. Annual precipitation provides some indication of regional change, but is not an adequate indicator when determining impacts on transportation as it masks much of the seasonal variability.

| Northeast (Δ Precipitation) | | Near-term (%) | Mid-century (%) | End-of-century (%) |
|---|---|---|---|---|
| **Winter** | Mean | 6 | 8 – 11 | 11 – 17 |
| | Likely | 2 – 11 | 2 – 18 | 4 – 27 |
| | Very Likely | (2) – 15 | (4) – 26 | (4) – 36 |
| **Spring** | Mean | 3 | 5 – 6 | 9 – 11 |
| | Likely | (2) – 7 | 0 – 12 | 1 – 21 |
| | Very Likely | (7) – 12 | (5) – 17 | (9) – 31 |
| **Summer** | Mean | 2 | 1 – 2 | (1) – 2 |
| | Likely | (1) – 6 | (6) – 7 | (12) – 11 |
| | Very Likely | (5) – 10 | (12) – 14 | (24) – 23 |
| **Fall** | Mean | 1 – 2 | 3 | 3 – 4 |
| | Likely | (4) – 6 | (3) – 9 | (5) – 13 |
| | Very Likely | (10) – 11 | (9) – 16 | (15) – 23 |

Table 3-5: Seasonal precipitation percent changes for the Northeast region over the *near-term* (2010-2029), *mid-century* (2040-2059) and *end-of-century* (2080-2098) relative to 1961-1979. The range values are from low (B1) and high (A2) emissions scenarios. Values in parentheses are negative values and represent decreases in precipitation. Data are from the USGCRP (2009).

### 3.3.2.2 Mid-century (2040-2070)

Overall, precipitation in the Northeast is projected to increase by *mid-century* across all seasons, with the greatest change again projected to occur in the winter months. By *mid-century*, the average winter precipitation increase projected for the Northeast is 8% and 11%, with a likely range of +2 to +18% (USGCRP 2009). Another study that used a higher bounding emission scenario (A1Fi) reported an even greater increase in winter precipitation of 11 to 16% for most of this region including New England, New York, New Jersey, and Pennsylvania (NECIA 2006). Spring and fall precipitation are projected to increase more moderately at 5 to 6% with a likely range of +0 to +12%, and 3% with a likely range of -3 to +9%, respectively. Summer precipitation is projected to increase the least of all the seasons, with an increase of only 1 to 2% and a likely range of -6 to +7% (USGCRP 2009).

By *mid-century*, this region is projected to experience more than an 8% increase in the average amount of rain that falls on any given rainy day, with the duration of extreme rain events[31] expected to increase by 1 to 1.5 days. In addition, the maximum amount of precipitation falling during any five-day period in a year is projected to increase by 8 to 13% (NECIA 2006). These increases in heavy rainfall events will increase the risk of floods for the Northeast region. NECIA (2006) suggests little change in the frequency of winter-time storms for the East Coast. However, under the "high-end" scenario (A1Fi), between 5 and 15% of these storms (an additional 1 storm per year) will move northward during late winter (Jan, Feb, March), affecting the Northeast. No change is projected for

---

[31] Extreme rain event is defined as more than 2 inches per day.

the "low-end" (B1) scenario. In addition, the impact of a higher sea level will increase the likelihood of storm damage to coastal locations.

### 3.3.2.3 End-of-century (2070-2100)

By the *end-of-century*, the Northeast is projected to experience the greatest seasonal increase in precipitation during the winter months. This increase is projected to be 11 to 17%, with a likely range of +4 to +27%. Summer will continue to be the least affected of the seasons, with average increases in total seasonal precipitation projected to be 2% under the high emissions scenario and -1% under the low emissions scenario (USGCRP 2009). By the end of the century, the intensity of any particular precipitation event is projected to increase, on average, by 12 to 13% (NECIA 2006). Additionally, the number of days in a given year with precipitation events of greater than two inches per day is projected to slightly increase by an additional 1.25 to 1.75 days per year (NECIA 2006). As air temperatures rise, the Northeast can expect a continuation of recent trends in the type of precipitation experienced during winter: less snow and more rain (NECIA 2006). Most of the Northeast could lose approximately four to 15 snow-covered days per winter month, with a 25 to 50% reduction in the length of the snow season with the onset of an earlier spring (NECIA 2006).

### 3.3.3 Sea-Level Rise

Global sea-level rise (SLR) of 7 to 79" (18cm to 2.0m) is projected for 2100 (see section 3.1.3. for discussion on global and local sea-level rise). SLR at the local/regional level is influenced by multiple factors, including sedimentation and erosion, ocean circulation, gravitationally induced changes, ocean density (affected by regional changes in ocean salinity and ocean temperature), and vertical motion of the land (subsidence or uplift). In the $20^{th}$ century, the relative sea-level rise for the Northeast was greater than the level of global sea-level rise.[32]

The following discussion describes studies providing local sea-level rise projections for the Northeast. As noted in section 3.1.3 above, making local or regional projections is highly uncertain, given the incomplete understanding of some of the effects that can take place at the local level. There is no study that considers all these factors, nor is there a consistent methodology for projecting sea-level rise applied across these studies. Therefore, local SLR projections, while informative, should be considered carefully, and with a clear understanding of what factors each study includes or excludes.

Yin et al. (2009) projects that the Northeast coastline could experience a sea-level rise much greater than the global average, due to changes in ocean circulation.[33] Yin et al.

---

[32] The National Water Level Observation provides historic sea level trends for 28 stations along the Northeast U.S. coastline. These measurements are provided by NOAA and include stations that provide sea level trends over a 30-year span or, in most cases, much longer (NOAA 2010). The Northeast trend is compared against the global observed trend of sea-level rise of $1.8 \pm 0.5$ mm yr$^{-1}$ from 1961 to 2003 (IPCC 2007a).

[33] Dynamic sea-level rise at New York City at *end-of-century* (i.e., 2091-2100) is projected using the GFDL CM2.1 climate model under B1 (low emission) and A2 (high emission) scenarios relative to 1981-2000 mean sea level (Yin et al. 2009).

(2009) estimate that these ocean circulation changes will increase sea level in New York City, NY by 5.9 to 8.3" (15cm to 21cm) above what would be expected from global sea-level rise and local vertical land motions alone (this study does not consider other factors such as erosion or sedimentation). Similar amplifications are projected for Washington, DC and Boston, MA.

Sea-level rise exacerbates the impacts of strong storm events. Kirshen et al.'s (2008) analysis indicates that the storm surge elevations across the Northeast associated with a storm that has a 1% chance of occurring in 2005 are projected to increase substantially due to sea-level rise, even as storm intensity remains unchanged.[34] Using the lower emission scenario (B1) as the lower bound and the Rahmstorf (2007) study as the upper bound, Kirshen et al. made the following projections for sea-level-rise-induced increases in storm surge elevation by 2100: Atlantic City, NJ is projected to experience the greatest increase of 46.9 to 74.4" (119cm to 189cm), and Woods Hole, MA is projected to experience the least at 13.2 to 28.8" (33.5 to 73.2cm). New York City, NY; Boston, MA; and New London, CT fell within the range between Woods Hole, MA and Atlantic City, NJ. This study further suggests that a flood that has a 1% chance of occurring in New York City in 2005 has about a 5% (low emission scenario) to a 50% chance (high emission scenario) of occurring in a year by 2100. Kirshen et al. (2008) did not consider other regional sea-level rise effects, such as ocean circulation or wind patterns, changes in the relative elevation of coastal land (i.e., caused by subsidence or uplift), or local changes in ocean density.

---

[34] Storm surge elevation along the Northeast coast at *end-of-century* (i.e., 2100) is projected using a long-term average of highest daily tides at each location, a mid-range of global sea-level rise predicted by IPCC under the B1 (low emissions) scenario, and high-range associated with the mid-range projection provided by Rahmstorf (2007), relative to the North American Vertical Datum (NAVD) in 1988. Tide measurements, due to the location of the tide gauges, measure both storm surge and increased river flow during coastal flooding events. (Kirshen et al. 2008).

## 3.4 Southeast

### 3.4.1 Temperature

#### 3.4.1.1 Near-term (2010-2040)

Within the next two decades, the annual mean temperature in the Southeast is projected to increase by approximately 2°F with a likely range of 1.7 to 2.7°F (USGCRP 2009; CCSP 2008a). This projected warming is greater than the 1.2°F increase already experienced over the 1993 to 2008 time period compared with a 1961-1979 baseline (USGCRP 2009).[35] Projected seasonal temperatures exhibit similar increases, with both summer and fall temperatures projected to increase by slightly greater amounts, and the winter and spring by slightly less (USGCRP 2009). Not only are summer temperatures projected to increase, but so are the number of extreme heat days.[36] By 2030, Houston, TX is projected under a higher emission scenario (A2) to experience a 25 to 75% probability of having 4 to 11 days above 100°F (in 2007, the probability of 4 days at 100°F was over 45% and less than 10% for 11 days) (CCSP 2008a).[37]

#### 3.4.1.2 Mid-century (2040-2070)

By *mid-century*, the increase in annual mean temperature is projected to be approximately 3.2 to 4.0°F, with a likely range of 2.4 to 4.8°F (USGCRP 2009). The projections for temperature increases in spring and fall are relatively similar to the annual mean. The mean temperature for the summer months is projected to be the highest mean increase of all the seasons, with a mean warming of 3.5 to 4.5°F and a likely range of 2.5 to 5.6°F. The mean temperature for the winter months is projected to be slightly lower than the annual mean, with a mean warming of 2.7 to 3.6°F and a likely range of 1.6 to 4.5°F (USGCRP 2009). Extreme heat days are projected to continue to increase. By 2060, Houston, TX is projected to have a 25 to 75% probability of having 14 to more than 20 days above 100°F per year (in 2007, the probability of 14 days or more at 100°F was less than 5%) (CCSP 2008a).

#### 3.4.1.3 End-of-century (2070-2100)

By the *end-of-century*, annual mean temperature in the Southeast region is projected to increase by 4.5 to 7.8°F, with a likely range of 3.4 to 9.4°F (USGCRP 2009). The projections for the spring months are similar to the annual mean. The Southeast is projected to experience the smallest warming in winter, with temperature increases of 4.0 to 6.3°F and a likely range of 2.8 to 7.9°F (USGCRP 2009). The summer months are projected to experience the greatest warming of all the seasons, with temperature

---

[35] This information was provided by personal communication with Jay H. Lawrimore of the National Climatic Data Center.
[36] Extreme heat day defined here as a daily maximum temperature above 100°F.
[37] CCSP 2008a study draws from results of 17 climate models represented in the CMIP3 dataset for IPCC AR4.

increases of 4.8 to 9.0°F and a likely range of 3.5 to 11.2°F (USGCRP 2009). The fall months are projected to display a slightly higher mean than the annual average, with mean increases projected to be 4.7 to 8.3°F and a likely range of 3.5 to 9.8°F (USGCRP 2009).

| Southeast (Δ Temperature) | | Near-term (°F) | Mid-century (°F) | End-of-century (°F) |
|---|---|---|---|---|
| Annual | Mean | 2.1 – 2.2 | 3.2 – 4.0 | 4.5 – 7.8 |
| | Likely | 1.7 – 2.7 | 2.4 – 4.8 | 3.4 – 9.4 |
| | Very Likely | 1.2 – 3.2 | 1.6 – 5.5 | 2.4 – 10.9 |
| Winter | Mean | 1.9 – 2.1 | 2.7 – 3.6 | 4.0 – 6.3 |
| | Likely | 1.1 – 2.8 | 1.6 – 4.5 | 2.8 – 7.9 |
| | Very Likely | 0.3 – 3.6 | 0.5 – 5.4 | 1.7 – 9.4 |
| Spring | Mean | 1.8 – 2.0 | 3.1 – 3.8 | 4.4 – 7.5 |
| | Likely | 1.3 – 2.7 | 2.2 – 4.6 | 3.2 – 9.1 |
| | Very Likely | 0.6 – 3.3 | 1.3 – 5.4 | 2.0 – 10.7 |
| Summer | Mean | 2.3 – 2.4 | 3.5 – 4.5 | 4.8 – 9.0 |
| | Likely | 1.5 – 3.0 | 2.5 – 5.6 | 3.5 – 11.2 |
| | Very Likely | 0.7 – 3.8 | 1.6 – 6.7 | 2.3 – 13.5 |
| Fall | Mean | 2.3 | 3.4 – 4.3 | 4.7 – 8.3 |
| | Likely | 1.8 – 2.9 | 2.6 – 4.9 | 3.5 – 9.8 |
| | Very Likely | 1.2 – 3.4 | 1.8 – 5.6 | 2.4 – 11.3 |

Table 3-6: Annual and seasonal temperature changes for the Southeast region over the *near-term* (2010-2029), *mid-century* (2040-2059) and *end-of-century* (2080-2098) relative to 1961-1979. The range values are from low (B1) and high (A2) emissions scenarios. Data are from the USGCRP (2009).

By the *end-of-century*, studies agree that the Southeast region will experience more extreme heat events. Between 2080 and 2100, extreme heat events (a combination of temperature and humidity) that currently have a 5% chance occurring for a given year are projected to have about a 50 to 100% chance of occurring each year (USGCRP 2009). By 2099, Houston, TX is projected to experience a near 100% probability of having more than 20 days above 100°F per year (current probability of 20 days at or above 100°F is near 0%) (CCSP 2008a). Similarly, the Appalachian Mountain region is predicted to experience approximately three times more "high temperature" days[38] each year by the *end-of-century* under a higher emission scenario (A2). (Diffenbaugh et al 2005).

### 3.4.2 Precipitation and Storm Events

#### 3.4.2.1 *Near-term* (2010-2040)

---

[38] High temperature days are defined as defined as being at or above the 95th percentile among current daily temperature records. This study uses RegCM3 and a baseline period of 1961-1985.

Current observations averaged over 1993 to 2008 suggest annual mean precipitation has decreased by 1% relative to 1961-1979 (USGCRP 2009).[39] In the *near-term*, mean precipitation in the Southeast does not exhibit a strong trend, but is generally projected to decrease in the summer and spring and increase in the fall. There is considerable disagreement between various climate models on the magnitude and, in some cases, direction of changes in precipitation in each time period.

Over the next two decades, the greatest increases in mean precipitation are projected to occur during the fall months and the greatest decreases during the spring months. The fall months are projected to increase by 1 to 2% with a likely range of -4 to +7%, while the spring months are projected to decrease by 0 to 2% with a likely range of -7 to +4% (USGCRP 2009). The winter and summer months exhibit very little change with a likely range of -6 to +5% and -8 to +8%, respectively (USGCRP 2009). In addition, if current observational trends continue, the spacing between precipitation events could continue to increase, leading to continued periods of drought (USGCRP 2009).

| Southeast (Δ Precipitation) | | *Near-term* (%) | *Mid-century* (%) | *End-of-century* (%) |
|---|---|---|---|---|
| **Winter** | Mean | (1) – 0 | (2) – 1 | (3) – 0 |
|  | Likely | (6) – 5 | (8) – 9 | (15) – 10 |
|  | Very Likely | (11) – 9 | (15) – 16 | (28) – 22 |
| **Spring** | Mean | (2) – 0 | 1 – 2 | (7) – 1 |
|  | Likely | (7) – 4 | (5) – 8 | (20) – 7 |
|  | Very Likely | (12) – 8 | (11) – 14 | (32) – 18 |
| **Summer** | Mean | 0 | (2) – 0 | (8) – 0 |
|  | Likely | (8) – 8 | (14) – 10 | (29) – 14 |
|  | Very Likely | (16) – 16 | (26) – 23 | (50) – 35 |
| **Fall** | Mean | 1 – 2 | (2) – (1) | 2 – 3 |
|  | Likely | (4) – 7 | (9) – 5 | (9) – 16 |
|  | Very Likely | (10) – 12 | (16) – 12 | (21) – 28 |

Table 3-7: Seasonal precipitation percent changes for the Southeast region over the *near-term* (2010-2029), *mid-century* (2040-2059) and *end-of-century* (2080-2098) relative to 1961-1979. The range values are from low (B1) and high (A2) emissions scenarios. Values in parentheses are negative values and represent decreases in precipitation. Data are from the USGCRP (2009).

3.4.2.2 Mid-century (2040-2070)

By *mid-century*, the greatest seasonal mean change in precipitation between emission scenario results is projected to occur during the winter months, with the mean ranging from an increase of 1% to a decrease of 2%, and a likely range of -8 to +9% (USGCRP 2009). The summer months exhibit a reduction in mean precipitation between 0 to 2%

---

[39] This information was provided by personal communication with Jay H. Lawrimore of the National Climatic Data Center. Annual precipitation provides some indication of regional change but is not an adequate indicator when determining impacts on transportation as it masks much of the seasonal variability.

with a likely range of -14 to +10%. The fall months are projected to sustain a reduction in mean precipitation of 1 to 2% with a likely range of -9 to +5% (USGCRP 2009). Spring is the only season projected to have an increase in precipitation with a mean change of 1 to 2%, and a likely range of -5 to +8% (USGCRP 2009).

### 3.4.2.3 End-of-century (2070-2100)

At the *end-of-century*, the largest seasonal mean decrease of precipitation is projected to occur during the summer season, with a mean decrease of about 0 to 8% and a likely range of -29 to +14% (USGCRP 2009). The spring season mean change ranges varies from 1 to -7% with a likely range of -20 to +7% (USGCRP 2009). The mean change for the winter season varies between 0% to a reduction of 3% with a likely range of -15 to +10% (USGCRP 2009). In contrast, the mean fall precipitation is projected to increase from 2 to 3% with a likely range -9 to +16% (USGCRP 2009). The greatest degree of certainty is that precipitation in the fall will increase, while the other precipitation effects are not as certain (USGCRP 2009). Diffenbaugh et al. (2005) found that the mid-Atlantic coast would be up to 40% rainier at the *end-of-century* overall.

By late century, the Gulf Coast region from Galveston, TX to Mobile Bay, AL is projected to experience an intensity increase of 5% for category 1 storms and 20% for category 4 storms (CCSP 2008a). This projection assumes that changes in hurricane intensity are directly related to increases in projected sea surface temperatures. For example, sea surface temperatures in the Atlantic hurricane formation region are projected to increase from 3 to 7°F, leading to increased tropical storms (CCSP 2008a). This does not take into account changes in other contributors to tropical storm development (e.g., vertical wind shear and vertical temperature structure), and hence, higher sea surface temperatures do not necessarily translate to an increased storm intensity.

## 3.4.3 Sea-Level Rise

Global sea-level rise (SLR) of 7 to 79" (18 cm to 2.0 m) is projected for 2100 (see section 3.1.3. for discussion on global and local sea-level rise). SLR at the local/regional level is influenced by multiple factors, including: sedimentation and erosion, ocean circulation, gravitationally induced changes, ocean density (affected by regional changes in ocean salinity and ocean temperature), and vertical motion of the land (subsidence/uplift). Using historical records, the relative sea-level rise for the Southeast was greater than the level of global sea-level rise.[40] This finding is consistent with the CCSP (2008a) study, which provides estimates of subsidence rates along the Gulf Coast that would lead to higher sea levels than the global average: Louisiana-Texas Chenier Plain at 0.19 in/yr

---

[40] The National Water Level Observation provides historic sea level trends for 37 stations along the Southeast U.S. coastline. These measurements are provided by NOAA and include stations that provide sea level trends over a 30-year span or, in most cases, much longer (NOAA 2010). The Southeast trend is compared against the global observed trend of sea-level rise of $1.8 \pm 0.5$ mm yr$^{-1}$ from 1961 to 2003 (IPCC 2007a).

(4.7 mm/yr), Louisiana Deltaic Plain at 0.32 in/yr (8.05 mm/yr), and Mississippi-Alabama Sound at 0.01 in/yr (0.34 mm/yr).

## 3.5 Midwest

### 3.5.1 Temperature

#### 3.5.1.1 Near-term (2010-2040)

Within the next two decades, the annual mean temperature in the Midwest is projected to increase by approximately 2.7°F with a likely range of 1.9 to 3.3°F (USGCRP 2009; Union of Concerned Scientists 2009). This projected warming is greater than the 1.4°F increase already experienced over the 1993 to 2008 time period compared with a 1961-1979 baseline (USGCRP 2009).[41] Projected summer, fall, and winter seasonal temperatures are expected to exhibit similar increases compared with the annual mean (USGCRP 2009). Spring seasonal temperatures are projected to increase slightly less than the annual mean at 2.0 to 2.4°F, with a likely range of 1.2 to 3.3°F (USGCRP 2009). These projections are similar to those provided for Chicago, IL, where temperature increases of approximately 1.8 to 3.6°F (1 to 2°C) are projected (Hellmann et al 2007).[42] Observed increases in the winter season have extended the frost-free (or growing) season by a week (USGCRP 2009); the growing season is expected to continue to lengthen as temperatures continue to warm in spring and fall. In addition, higher temperatures lead to increased evaporation, reducing water levels in the Great Lakes (USGCRP 2009).

#### 3.5.1.2 Mid-century (2040-2070)

By *mid-century*, the annual mean temperature increase is projected to be approximately 4.0 to 5.0°F, with a likely range of 3.0 to 6.0°F (USGCRP 2009). The projections for summer, fall, and winter temperatures are again similar to the annual mean. However, the temperature increase for the spring months is projected to be less, with a seasonal average of 3.6 to 4.2°F and a likely range of 2.2 to 5.6°F (USGCRP 2009). Chicago may experience a higher annual temperature than that suggested for the region, with an increase between 2.7 to 9°F (1.5 to 5°C) by 2070 (Hellmann et al 2007).

#### 3.5.1.3 End-of-century (2070-2100)

By the *end-of-century*, annual mean temperature is projected to increase by approximately 5.6 to 9.6°F in the Midwest with a likely range of 4.3 to 11.7°F (USGCRP 2009). The projections for fall and winter temperatures continue to be similar to the annual mean. However, the spring months are projected to experience a smaller temperature increase of 5.1 to 8.4°F, with a likely range of 3.5 to 10.6°F (USGCRP 2009). The summer months display a higher annual mean temperature increase of 5.6 to 10.8°F, with a likely range of 4.2 to 14.2°F (USGCRP 2009). Annual mean temperature

---

[41] This information was provided by personal communication with Jay H. Lawrimore of National Climatic Data Center.
[42] Hellman et al (2007) study projects changes in annual average temperature relative to 1961-1990 drawing from 21 IPCC AR4 models using a higher (A1Fi) emission scenario and lower (B1) emission scenario. The projection provided is for 2010-2039.

increases in Chicago are projected to be similar to these regional increases (Hellmann et al 2007). However, one set of studies found even higher annual mean temperature increases ranging between 6 and 14°F in Minnesota, Missouri, Indiana, Wisconsin, and Ohio, associated with a lower (B1) emission scenario and a higher (A1Fi) emission scenario (Union of Concerned Scientists 2009a-e).[43]

| Midwest (Δ Temperature) | | Near-term (°F) | Mid-century (°F) | End-of-century (°F) |
|---|---|---|---|---|
| Annual | Mean | 2.6 – 2.7 | 4.0 – 5.0 | 5.6 – 9.6 |
| | Likely | 1.9 – 3.3 | 3.0 – 6.0 | 4.3 – 11.7 |
| | Very Likely | 1.3 – 3.9 | 1.9 – 7.0 | 3.0 – 13.8 |
| Winter | Mean | 2.6 – 3.0 | 4.1 – 5.3 | 6.0 – 9.4 |
| | Likely | 1.6 – 4.0 | 2.9 – 6.6 | 4.6 – 11.5 |
| | Very Likely | 0.6 – 4.9 | 1.7 – 7.9 | 3.3 – 13.5 |
| Spring | Mean | 2.0 – 2.4 | 3.6 – 4.2 | 5.1 – 8.4 |
| | Likely | 1.2 – 3.3 | 2.2 – 5.6 | 3.5 – 10.6 |
| | Very Likely | 0.4 – 4.1 | 0.8 – 7.0 | 1.9 – 12.9 |
| Summer | Mean | 2.6 – 2.8 | 4.1 – 5.3 | 5.6 – 10.8 |
| | Likely | 1.9 – 3.8 | 2.8 – 6.8 | 4.2 – 14.2 |
| | Very Likely | 1.0 – 4.7 | 1.5 – 8.3 | 2.7 – 17.5 |
| Fall | Mean | 2.6 – 2.7 | 4.0 – 4.9 | 5.5 – 9.6 |
| | Likely | 2.0 – 3.4 | 2.9 – 5.8 | 4.1 – 11.6 |
| | Very Likely | 1.3 – 4.1 | 1.7 – 6.7 | 2.7 – 13.6 |

Table 3-8: Annual and seasonal temperature changes for the Midwest region over the *near-term* (2010-2029), *mid-century* (2040-2059) and *end-of-century* (2080-2098) relative to 1961-1979. The range values are from low (B1) and high (A2) emissions scenarios. Data are from the USGCRP (2009).

By the *end-of-century*, heat waves in the Midwest are expected to become longer, hotter, and more frequent (Ebi and Meehl 2007; USGCRP 2009). By 2100, under a higher (A2) emission scenario, a heat event that currently has a 5% chance of occurring for a given year is projected to have about a 50 to 100% chance of occurring for a given year (USGCRP 2009). In Chicago and Cincinnati, the frequency of heat waves is expected to increase 24% and 50% respectively under a business as usual scenario (Ebi and Meehl 2007).[44] The average duration of these heat waves is also projected to increase by approximately 20% in both cities. Similarly, under a high (A1Fi) emission scenario, cities across the Midwest, including Des Moines, Cincinnati, and Indianapolis, are projected to experience between 65 and 85 days over 90°F each summer by the *end-of-century*

---

[43] This study provides projections based on statistically downscaled data of three climate models (CM2.1, HadCM3, and PCM) which represent the spectrum of climate sensitivity; the baseline period is 1961-1990.
[44] This study defines a heat wave as the maximum temp exceeding the 97.5 percentile for at least 3 days, the average minimum temperature above the 97.5 percentile for at least 3 days, and the maximum temperature above the 81st percentile for the entire period. This study used a 'business as usual' scenario and 1961-1990 as relative baseline.

compared with approximately 10 to 20 days averaged over 1961-1990 (Union of Concerned Scientists 2009a).[45]

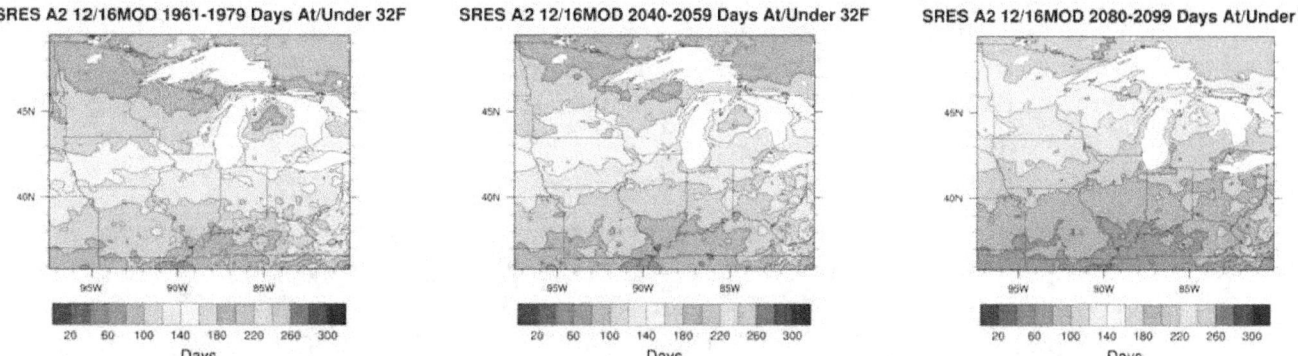

Figure 8: Average number of freezing days per year in the Midwest (defined as days the minimum temperature is below 32°F). The illustrations are 16 multi-model ensemble averages for years 1961-1979, 2040-2059, and 2080-2099 for SRES A2. (USGCRP 2009)

Figure 8 illustrates the reduction of freezing days projected for the Midwest; this is particularly evident for the northern areas. For example, southern Minnesota experienced about 170 freezing days in 1961-1979 and is projected to experience about 110 freezing days in 2080-2099 under the A2 scenario.

### 3.5.2 Precipitation and Storm Events

#### 3.5.2.1 Near-term (2010-2040)

Current observations averaged over 1993 to 2008 suggest that annual mean precipitation has increased by 5% relative to the 1961-1979 time period (USGCRP 2009).[46] By far the largest seasonal increase in precipitation is projected to occur during the winter months, with an average increase of 6 to 7% and a likely range of +2 to +12% (USGCRP 2009). Annual mean precipitation in Chicago is projected to experience precipitation increases in line with the regional estimates (Hellmann et al. 2007). Heavy precipitation events are also projected to increase during this time, with the frequency of spring rainfall heavy downpours increasing by almost 15% in Missouri, Illinois, and Minnesota under a high emission scenario (A1Fi) compared with 1961-1990 (Union of Concerned Scientists 2009a).[47] In the next two decades, heavy rains are projected to increase by 66% in St. Paul, 35% in Indianapolis, and 20% in Chicago (Union of Concerned Scientists 2009). These increases are expected to increase flooding and overload many drainage systems (USGCRP 2009).

---

[45] This study uses three state-of-the-art global climate models (CM 2.1, HadCM3, and PCM) that together adequately capture climate sensitivity.

[46] This information was provided by personal communication with Jay H. Lawrimore of the National Climatic Data Center. Annual precipitation provides some indication of regional change but is not an adequate indicator when determining impacts on transportation as it masks much of the seasonal variability.

[47] This study defines a heavy downpour as more than 2" of rain per day.

### 3.5.2.2 Mid-century (2040-2070)

Precipitation increases are projected to occur in the winter and spring months. Winter precipitation, for example, is projected to increase by 8 to 9% with a likely range of +1 to +15% (USGCRP 2009). Precipitation is projected to decrease by an average of 1 to 4% in the summer months with a likely range of -15 to +8% (USGCRP 2009). However, studies disagree on the magnitude of predicted changes in precipitation. Precipitation changes in Chicago are projected to be between -2 and 10% by 2070 (Hellmann et al 2007).

| Midwest ($\Delta$ Precipitation) | | Near-term (%) | Mid-century (%) | End-of-century (%) |
|---|---|---|---|---|
| **Winter** | Mean | 6 – 7 | 8 – 9 | 10 – 14 |
| | Likely | 2 – 12 | 1 – 15 | 3 – 22 |
| | Very Likely | (3) – 16 | (6) – 21 | (3) – 30 |
| **Spring** | Mean | 3 – 4 | 7 – 9 | 10 – 14 |
| | Likely | (1) – 8 | 3 – 13 | 2 – 25 |
| | Very Likely | (6) – 12 | (1) – 18 | (9) – 36 |
| **Summer** | Mean | (1) | (4) – (1) | (9) – (2) |
| | Likely | (7) – 6 | (15) – 8 | (31) – 14 |
| | Very Likely | (14) – 13 | (26) – 19 | (53) – 36 |
| **Fall** | Mean | 1 | 1 – 3 | 2 – 3 |
| | Likely | (5) – 7 | (6) – 11 | (10) – 17 |
| | Very Likely | (11) – 13 | (14) – 18 | (23) – 30 |

Table 3-9: Seasonal precipitation percent changes for the Midwest region over the *near-term* (2010-2029), *mid-century* (2040-2059) and *end-of-century* (2080-2098) relative to 1961-1979. The range values are from low (B1) and high (A2) emissions scenarios. Values in parentheses are negative values and represent decreases in precipitation. Data are from the USGCRP (2009).

### 3.5.2.3 End-of-century (2070-2100)

Again, the majority of the increase in precipitation will occur during the winter and spring months, in which precipitation is projected to increase by an average of 10 to 14% with a likely range of +3 to +22%, and 10 to 14% with a likely range of +2 to +25%, respectively (USGCRP 2009). Average summer precipitation is projected to decrease by 2 to 9% with a likely range of -31 to +14% (USGCRP 2009). USGCRP (2009) projects the likelihood of summer-time drought increasing. Annual mean precipitation in Chicago is projected to change by -1 to +19% by the *end-of-century* (Hellmann et al. 2007). There was some disagreement between studies on the magnitude of seasonal mean precipitation. A study by the Union of Concerned Scientists (2009) found higher overall increases in winter precipitation, with 20-50% increases in Missouri, Minnesota, and Michigan.[48] The

---

[48] This study uses three state-of-the-art global climate models (CM 2.1, HadCM3, and PCM) that together adequately capture climate sensitivity, and the B1 emission scenario (low) and A1Fi emission scenario (high).

same study found 10-20% less rain in summer precipitation for most of the Midwest region by the *end-of-century*. Heavy spring downpours (defined as two inches of rain in one day) are expected to become more frequent during this time frame, with approximately 30% increases projected for Iowa, Ohio, Illinois, and Wisconsin (Union of Concerned Scientists 2009a-e).

## 3.6 Great Plains

### 3.6.1 Temperature

#### 3.6.1.1 Near-term (2010-2040)

Within the next two decades, the annual mean temperature of the Great Plains is projected to increase by approximately 2.5°F, with a likely range of 1.8 to 3.1°F (USGCRP 2009). This projected warming is greater than the 1.3°F increase already experienced over the 1993 to 2008 time period relative to a 1961-1979 baseline (USGCRP 2009).[49] Fall temperature increases are projected to be similar to the projected annual mean warming. Summer temperature increases are projected to be slightly higher than the annual mean warming, at 2.7 to 2.9°F, with a likely range of 1.8 to 3.7°F (USGCRP 2009). Spring temperature increases are expected to be slightly lower than the annual mean increase, at 1.9 to 2.2°F, with a likely range of 1.2 to 3.0°F (USGCRP 2009); likewise, winter temperature increases are also expected to be slightly lower than the annual mean increase, at 2.2 to 2.5°F, with a likely range of 1.4 to 3.4°F (USGCRP 2009).

| Great Plains (Δ Temperature) | | Near-term (°F) | Mid-century (°F) | End-of-century (°F) |
|---|---|---|---|---|
| **Annual** | Mean | 2.4 – 2.5 | 3.8 – 4.7 | 5.4 – 9.2 |
| | Likely | 1.8 – 3.1 | 2.7 – 5.8 | 3.9 – 11.2 |
| | Very Likely | 1.1 – 3.8 | 1.6 – 6.9 | 2.5 – 13.2 |
| **Winter** | Mean | 2.2 – 2.5 | 3.6 – 4.3 | 5.3 – 8.3 |
| | Likely | 1.4 – 3.4 | 2.4 – 5.6 | 3.8 – 10.4 |
| | Very Likely | 0.6 – 4.2 | 1.2 – 6.9 | 2.2 – 12.5 |
| **Spring** | Mean | 1.9 – 2.2 | 3.4 – 4.0 | 4.8 – 8.0 |
| | Likely | 1.2 – 3.0 | 2.1 – 5.5 | 3.1 – 10.3 |
| | Very Likely | 0.5 – 3.9 | 0.8 – 6.9 | 1.3 – 12.7 |
| **Summer** | Mean | 2.7 – 2.9 | 4.3 – 5.6 | 5.8 – 10.6 |
| | Likely | 1.8 – 3.7 | 3.0 – 7.1 | 4.1 – 13.6 |
| | Very Likely | 0.8 – 4.6 | 1.7 – 8.7 | 2.4 – 16.6 |
| **Fall** | Mean | 2.4 – 2.5 | 3.8 – 4.7 | 5.5 – 9.6 |
| | Likely | 1.8 – 3.3 | 2.7 – 5.7 | 4.0 – 11.5 |
| | Very Likely | 1.1 – 4.0 | 1.6 – 6.7 | 2.4 – 13.5 |

Table 3-10: Annual and seasonal temperature changes for the Great Plains region over the *near-term* (2010-2029), *mid-century* (2040-2059) and *end-of-century* (2080-2098) relative to 1961-1979. The range values are from low (B1) and high (A2) emissions scenarios. Data are from the USGCRP (2009).

---

[49] This information was provided by personal communication with Jay H. Lawrimore of the National Climatic Data Center.

### 3.6.1.2 Mid-century (2040-2070)

By *mid-century*, annual mean temperature is projected to increase by 3.8 to 4.7°F with a likely range of 2.7 to 5.8°F (USGCRP 2009). Fall temperatures are expected to increase similarly to the annual mean. Summer months are projected to have a greater temperature increase of 4.3 to 5.6°F, with a likely range of 3.0 to 7.1°F (USGCRP 2009). Winter warming is projected to be lower than the annual mean increase. Warming in the spring is projected to be the least of all seasons, with an increase of 3.4 to 4.0°F and a likely range of 2.1 to 5.5°F (USGCRP 2009).

### 3.6.1.3 End-of-century (2070-2100)

By the *end-of-century*, annual mean temperature is projected to have increased by 5.4°F to 9.2°F with a likely range of 3.9 to 11.2°F (USGCRP 2009). Seasonal warming trends are projected to continue, with slightly greater or similar warming in the summer and fall months, and less warming in the winter and spring months. Summer mean temperatures are expected to increase 5.8 to 10.6°F, with a likely range of 4.1 to 13.6°F (USGCRP 2009). The smallest seasonal temperature increase is projected to occur in the spring, with an increase of 4.8 to 8.0°F and a likely range of 3.1 to 10.3°F (USGCRP 2009).

Lenihan et al. (2008) made projections of changes in *maximum* temperatures for the Great Plains region (the USGCRP projections reported above are for mean temperature).[50] They projected that the late century increase in maximum temperatures would be greatest in the central and northern areas of the Great Plains. The authors found that average monthly maximum temperatures would increase by 7 to 13°F (4 to 7°C) across Oklahoma, Kansas, Nebraska, and North and South Dakota under lower emission (B2) and higher emission (A2) scenarios. Texas displayed smaller increases ranging from 5 to 9°F (3 to 5°C).

## 3.6.2 Precipitation and Storm Events

### 3.6.2.1 Near-term (2010-2040)

Averaging the 1993 to 2008 time period of observations across the Great Plains region suggests that annual mean precipitation has increased by 4% relative to the 1961-1979 time period (USGCRP 2009).[51] In the *near-term*, mean precipitation is generally projected to increase in the winter and spring, and to decrease in the summer. It is unclear whether fall precipitation will increase or decrease. There is considerable disagreement between various climate models on the magnitude and direction of changes in precipitation in that season.

---

[50] This study provides results averaged across three GCMs for 2070-2099 relative to a 1971 to 2000 baseline.
[51] This information was provided by personal communication with Jay H. Lawrimore of the National Climatic Data Center. Annual precipitation provides some indication of regional change but is not an adequate indicator when determining impacts on transportation as it masks much of the seasonal variability.

Over the next two decades, mean precipitation is projected to increase by 3% in the winter, with a likely range of -2 to +7%. The spring increase is projected to be between 1 to 2%, with a likely range of -3 to +6% (USGCRP 2009). Precipitation may decrease in the summer by 2 to 3%, with a likely range of -9 to +4%. The likely range for precipitation in the fall is between -5 and +6% (USGCRP 2009).

### 3.6.2.2 Mid-century (2040-2070)

Similar trends in precipitation are projected for the Great Plains through the middle of the century. The winter months are projected to experience a mean precipitation increase of 4 to 5%, with a likely range of -1 to +9%. An increase of roughly 3% is projected for the spring, with a likely range of -3 to +8% (USGCRP 2009). Precipitation is projected to decline by 3 to 5% in the summer with a likely range of -18 to +7%. The likely range of precipitation change in the fall is -9 to +7% (USGCRP 2009).

| Great Plains (Δ Precipitation) | | Near-term (%) | Mid-century (%) | End-of-century (%) |
|---|---|---|---|---|
| Winter | Mean | 3 | 4 – 5 | 5 – 8 |
| | Likely | (2) – 7 | (1) – 9 | (1) – 17 |
| | Very Likely | (6) – 11 | (6) – 14 | (9) – 25 |
| Spring | Mean | 1 – 2 | 3 | 3 – 4 |
| | Likely | (3) – 6 | (3) – 8 | (7) – 12 |
| | Very Likely | (7) – 9 | (9) – 14 | (16) – 21 |
| Summer | Mean | (3) – (2) | (5) – (3) | (9) – (3) |
| | Likely | (9) – 4 | (18) – 7 | (29) – 11 |
| | Very Likely | (15) – 11 | (30) – 20 | (49) – 31 |
| Fall | Mean | 0 | (1) | (1) – 2 |
| | Likely | (5) – 6 | (9) – 7 | (17) – 12 |
| | Very Likely | (11) – 12 | (17) – 14 | (31) – 26 |

Table 3-11: Seasonal precipitation percent changes for the Great Plains region over the *near-term* (2010-2029), *mid-century* (2040-2059) and *end-of-century* (2080-2098) relative to 1961-1979. The range values are from low (B1) and high (A2) emissions scenarios. Values in parentheses are negative values and represent decreases in precipitation. Data are from the USGCRP (2009).

### 3.6.2.3 End-of-century (2070-2100)

By late century, projections indicate that the Great Plains are expected to continue to experience wetter winters and springs, and drier summers. Mean precipitation is projected to increase by 5 to 8% in the winter, with a likely range of -1 to +17%. Spring precipitation is projected to increase by 3 to 4%, with a likely range of -7 to +12% (USGCRP 2009). The summer months are projected to have 3 to 9% lower mean precipitation, with a likely range of -29 to +11%. Projections of fall precipitation range from a 1% decrease to a 2% increase, with a likely range of -17 to +12% (USGCRP 2009).

## 3.7 Southwest

### 3.7.1 Temperature

#### 3.7.1.1 Near-term (2010-2040)

Within the next several decades, the Southwest can expect to see increases in both annual average and seasonal average temperatures, with the greatest warming expected in the summer months. The projection for the annual mean temperature increase is approximately 2.4°F, with a likely range of 1.7 to 3.0°F; the fall months are projected to warm similarly (USGCRP 2009). This projected warming is greater than the 1.6°F increase already experienced over the 1993 to 2008 time period relative to a 1961-1979 baseline (USGCRP 2009).[52] The summer months are projected to experience the greatest warming, with an increase of approximately 2.7°F and a likely range of 1.8 to 3.4°F (USGCRP 2009). The spring and winter months are projected to have the smallest mean temperature increase, approximately 2.2°F, and a likely range of 1.3 to 3.1°F (USGCRP 2009).

#### 3.7.1.2 Mid-century (2040-2070)

By *mid-century*, the ranges in mean temperature projections widen as a result of the widening of plausible scenarios. The annual mean temperature increase for the Southwest is projected to be 3.6 to 4.5°F with a likely range of 2.6 to 5.5°F (USGCRP 2009). Spring and fall season averages are projected to change similarly to the annual mean. Winter is expected to see less substantial increases than the other seasons, with projections of 3.2 to 3.9°F and a likely range of 2.0 to 5.1°F (USGCRP 2009). Conversely, *mid-century* summer temperatures are projected to continue to warm more than other seasons, with mean temperature increases of 4.1 to 5.3°F and a likely range of 3.1 to 6.5 °F (USGCRP 2009). Cayan et al. (2009) project annual mean temperature increases for the state of California in 2050 to be between 1.8 to 5.4°F (1 to 3°C) under the lower (B1) and higher (A2) emission scenarios, a much greater range than those provided for the Southwest.[53]

#### 3.7.1.3 End-of-century (2070-2100)

By the *end-of-century*, the annual mean temperature in the Southwest region is projected to increase considerably compared with the earlier time horizons. The annual mean temperature increase for the region is projected to be 5.2 to 8.7°F, with a likely range of 3.8 to 10.2°F (USGCRP 2009). As with previous periods, summer is projected to experience the greater warming, with temperature increases of 5.6 to 9.7°F and a likely range of 4.2 to 11.6°F. Winter is projected to see the smallest seasonal temperature increase of 4.8 to 7.6°F, with a likely range of 3.3 to 9.4°F (USGCRP 2009). Projections

---

[52] This information was provided by personal communication with Jay H. Lawrimore of the National Climatic Data Center.
[53] This study uses results downscaled from six climate models (CM3, CM2.1, MICRO3.2, ECHAM5, CCSM3, PCM1) for mid-century versus a 1961 to 1990 baseline.

from Cayan et al. (2009) for California are qualitatively consistent with the USGCRP (2009) results. They project the annual mean temperature increase for the state of California to be roughly 4 to 9°F (2 to 5°C) under the lower (B1) and higher (A2) emission scenarios, with the smallest projected warming in the winter season of 1.8 to 7.2°F (1 to 4°C) and the greatest projected warming in the summer season of 2.7 to 10.8°F (1.5 to 6°C).

| Southwest (Δ Temperature) | | Near-term (°F) | Mid-century (°F) | End-of-century (°F) |
|---|---|---|---|---|
| Annual | Mean | 2.3 – 2.4 | 3.6 – 4.5 | 5.2 – 8.7 |
| | Likely | 1.7 – 3.0 | 2.6 – 5.5 | 3.8 – 10.2 |
| | Very Likely | 1.0 – 3.7 | 1.6 – 6.4 | 2.5 – 11.8 |
| Winter | Mean | 2.1 – 2.2 | 3.2 – 3.9 | 4.8 – 7.6 |
| | Likely | 1.4 – 3.0 | 2.0 – 5.1 | 3.3 – 9.4 |
| | Very Likely | 0.6 – 3.8 | 0.8 – 6.2 | 1.8 – 11.3 |
| Spring | Mean | 2.1 – 2.2 | 3.5 – 4.1 | 4.9 – 8.0 |
| | Likely | 1.3 – 3.1 | 2.1 – 5.2 | 3.3 – 9.9 |
| | Very Likely | 0.4 – 4.0 | 0.8 – 6.3 | 1.7 – 11.8 |
| Summer | Mean | 2.6 – 2.7 | 4.1 – 5.3 | 5.6 – 9.7 |
| | Likely | 1.8 – 3.4 | 3.1 – 6.5 | 4.2 – 11.6 |
| | Very Likely | 1.1 – 4.2 | 2.1 – 7.7 | 2.8 – 13.5 |
| Fall | Mean | 2.3 – 2.4 | 3.7 – 4.6 | 5.3 – 9.2 |
| | Likely | 1.7 – 3.0 | 2.8 – 5.4 | 4.0 – 10.6 |
| | Very Likely | 1.1 – 3.6 | 2.0 – 6.2 | 2.8 – 12.0 |

Table 3-12: Annual and seasonal temperature changes for the Southwest region over the *near-term* (2010-2029), *mid-century* (2040-2059) and *end-of-century* (2080-2098) relative to 1961-1979. The range values are from low (B1) and high (A2) emissions scenarios. Data are from the USGCRP (2009).

Diffenbaugh et al. (2005) project that the Southwest region could see up to 100 "high temperature" days[54] per year under a higher (A2) emission scenario. The frequency in high temperature days for California and Utah is projected to increase by four times today's numbers. Additionally, the USGCRP (2009) describes heat events (a combination of heat and humidity) that currently have a 5% chance of occurring for a given year are projected to increase in frequency, with about a 50 to 100% chance of occurring for a given year under a higher (A2) emission scenario. Figure 9 illustrates the geographic variability of experiencing heat days today across the Southwest and further demonstrates how this variability and frequency will increase by the end of the century.

---

[54] "High temperature" days are defined as being at or above the 95th percentile among current daily temperature records. This study uses a regional model (RegCM3) to provide projections in 2071-2095 relative to 1961-1985.

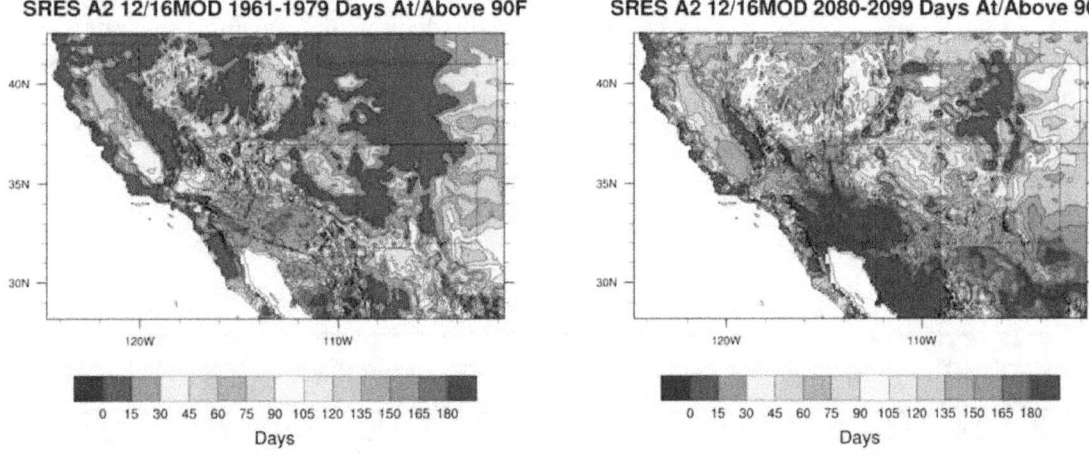

**Figure 9: Average number of heat days per year (defined as days the maximum temperature is above 90°F). The illustrations are 16 multi-model ensemble averages for years 1961-1979, and 2060-2099 for SRES A2. (USGCRP 2009)**

### 3.7.2 Precipitation and Storm Events

#### 3.7.2.1 Near-term (2010-2040)

Observations in the Southwest region averaged over 1993 to 2008 suggest that annual mean precipitation has increased by less than 1% relative to the 1961-1979 time period (USGCRP 2009).[55] Within the next several decades, winter precipitation is expected to increase while precipitation in the other three seasons is projected to decrease. Winter precipitation is projected to increase by 2 to 4%, with a likely range of -6 to +14% (USGCRP 2009). The greatest seasonal decrease in precipitation of 4 to 5% is projected for the summer season, with an associated likely range of -14 to +4% (USGCRP 2009).

#### 3.7.2.2 Mid-century (2040-2070)

The general precipitation trends projected for the next several decades are expected to continue through the middle of the century: winter will likely see increases in precipitation while the other seasons can expect decreases. Winter precipitation is projected to increase by 1 to 5%, with a likely range of -6 to +16% (USGCRP 2009). The summer and fall seasons are projected to experience decreases of 5 to 8% with a likely range of -22 to +7%; and 2 to 3% with a likely range of -11 to +5%, respectively (USGCRP 2009). The greatest decrease in precipitation of 6 to 10% is projected for spring, with a likely range of -20 to 0% (USGCRP 2009).

For transportation planning purposes, the *type* of precipitation is similar in importance to the *amount*. In the Southwest, the type of precipitation (rain vs. snow) will be affected by the earlier onset of spring following warmer winters. Leung et al. (2004) project that the

---

[55] This information was provided by personal communication with Jay H. Lawrimore of the National Climatic Data Center. Annual precipitation provides some indication of regional change; however, it is not an adequate indicator when determining impacts on transportation as it masks much of the seasonal variability.

Sierra Nevada Mountains will see a 60 to 70% decline in snowpack by *mid-century* averaged over 2040 to 2060 under a business as usual scenario.[56] During the same timeframe, more precipitation is likely to fall as rain than snow in the Colorado River Basin, with a 10 to 20% reduction in snow, and a more than 30% reduction in the Sacramento-San Joaquin River basin (Leung et al. 2004). Averaged over 2035 to 2064, the amount of water stored as snow in the Sierra Nevada Mountains as of April 1 is projected to decrease by 12 to 42% at all elevations under a lower (B1) and higher (A2) emission scenario (Cayan et al. 2008).[57]

| Southwest (Δ Precipitation) | | *Near-term* (%) | *Mid-century* (%) | *End-of-century* (%) |
|---|---|---|---|---|
| **Winter** | Mean | 2 – 4 | 1 – 5 | 2 – 5 |
| | Likely | (6) – 14 | (6) – 16 | (13) – 23 |
| | Very Likely | (16) – 24 | (17) – 27 | (30) – 40 |
| **Spring** | Mean | (5) – (4) | (10) – (6) | (19) – (7) |
| | Likely | (10) – 2 | (20) – 0 | (32) – 1 |
| | Very Likely | (16) – 8 | (31) – 10 | (45) – 9 |
| **Summer** | Mean | (5) – (4) | (8) – (5) | (5) – (3) |
| | Likely | (14) – 4 | (22) – 7 | (24) – 13 |
| | Very Likely | (23) – 13 | (36) – 21 | (43) – 32 |
| **Fall** | Mean | (1) – 0 | (3) – (2) | (1) – 0 |
| | Likely | (6) – 6 | (11) – 5 | (15) – 12 |
| | Very Likely | (12) – 11 | (20) – 13 | (28) – 26 |

Table 3-13: Seasonal precipitation percent changes for the Southwest region over the *near-term* (2010-2029), *mid-century* (2040-2059) and *end-of-century* (2080-2098) relative to 1961-1979. The range values are from low (B1) and high (A2) emissions scenarios. Values in parentheses are negative values and represent decreases in precipitation. Data are from the USGCRP (2009).

### 3.7.2.3 End-of-century (2070-2100)

By the *end-of-century*, winter precipitation is projected to increase by 2 to 5%, with a likely range of -13 to +23%, while the remaining seasons are projected to continue to decline in seasonal average precipitation (USGCRP 2009). The spring season is projected to experience the greatest decrease in precipitation of 7 to 19%, with a likely range of -32 to +1% (USGCRP 2009). It is expected that the shift in precipitation type from snow to rain will continue (USGCRP 2009). The amount of water stored as snow on April 1 is projected to decrease by 32 to 79% under a lower (B1) and higher (A2) emission scenario (Cayan et al. 2008). In sum, this region is expected to endure a greater likelihood of drought while, conversely, having an increased risk of flooding (USGCRP 2009).

---

[56] This study uses ensemble simulations of PCM/MM5.
[57] Three climate models (CM3, CM2.1, PCM2) are downscaled for this study and a reference period of 1961-1990 is used.

## 3.7.3 Sea-Level Rise

Global sea-level rise (SLR) of 7 to 79" (18cm to 2.0m) is projected for 2100 (see section 3.1.3 for discussion on global and local sea-level rise). SLR at the local/regional level is influenced by multiple factors, including sedimentation and erosion, ocean circulation, gravitationally induced changes, ocean density (affected by regional changes in ocean salinity and ocean temperature), and vertical motion of the land (subsidence or uplift). In the 20$^{th}$ century, the relative sea-level rise for the Southwest region is generally similar to the level of global sea-level rise.[58]

The following discussion describes studies providing local sea-level rise projections for the Southwest region. As noted in section 3.1.3 above, making local or regional projections is highly uncertain, given the incomplete understanding of some of the effects that can take place at the local level. There is no study that considers all these factors, nor is there a consistent methodology applied across these studies. Therefore, local SLR projections, while informative, should be considered carefully, and with a clear understanding of what factors each study includes or excludes.

Projections of local sea-level rise along the California coast have been developed that take into account tides, weather, and monthly and interannual sea level fluctuations from El Niño/Southern Oscillation (Cayan et al. 2008). These projections are not provided here as they do not consider other effects such as the changes in elevation of coastal land (i.e., caused by subsidence or uplift), local changes in ocean density, or erosion and sedimentation on local sea-level rise. These projections are driven by conservative IPCC global sea-level rise estimates that have been enhanced with accelerated estimates of projected ice melt.[59,60] The Climate Change Effects Typology Matrix provides the results of this study and further finds that if the sea-level rise realized was within a moderate estimate of the Table 3-1 projections, extreme events and their duration would increase substantially.

---

[58] The National Water Level Observation provides historic sea level trends for approximately 16 stations for the Southwest U.S. coastline. These measurements are provided by NOAA and include stations that provide sea level trends over a 30 year span or, in most cases, much longer (NOAA 2010). The Southwest trend is compared against the global observed trend of sea-level rise of $1.8 \pm 0.5$ mm yr$^{-1}$ from 1961 to 2003 (IPCC 2007a).

[59] Local sea-level rise along the coast of California by end of century (i.e., 2070-2099) for B1 (low emissions) and A1 (high emissions) scenarios, relative to 2000. (Cayan, 2006)

[60] Cayan (2008) projected relative sea-level rise at three locations along the coast of California, by mid century (i.e., 2035-2065), and at end of century (i.e., 2070-2099), using a tidal prediction program, projections of sea level pressure and offshore wind stresses from GFDL and PCM climate models, projections of El Niño/Southern Oscillation variability, and global sea-level rise projections. Relative sea level projections were made under B1 (low emissions) and A2 (high emissions) scenarios relative to 1961-1990.

## 3.8 Pacific Northwest

### 3.8.1 Temperature

#### 3.8.1.1 Near-term (2010-2040)

Within the next two decades, the annual mean temperature is projected to increase for the Pacific Northwest by approximately 2°F, with a likely range of 1°F to 3°F (USGCRP 2009; Mote et al. 2005; Mote and Salathe 2009). This projected warming is greater than the 1°F increase already experienced over the 1993 to 2008 time period relative to a 1961-1979 baseline (USGCRP 2009).[61] Winter and fall changes are expected to be relatively similar to the annual mean. Summer months are projected to experience the greatest warming of 2.5 to 2.8°F and a likely range of 1.6 to 3.7°F (USGCRP 2009), whereas spring months are projected to experience the smallest warming of 1.7 to 1.9°F and a likely range of 0.8 to 3.0°F (USGCRP 2009).

#### 3.8.1.2 Mid-century (2040-2070)

By *mid-century*, the annual mean temperature increase is projected to be approximately 3.6 to 4.3°F, with a likely range of 2.6 to 5.4°F (USGCRP 2009). For 2040, Mote et al. (2005) projected a slightly lower annual mean temperature increase of 3.0°F (1.6°C). The projections of warming during winter and fall are relatively similar to the annual mean. The smallest warming of seasonal mean temperature is projected to occur during the spring months, at 3.1 to 3.4°F and a likely range of 1.7 to 4.7°F (USGCRP 2009). The largest warming of seasonal mean temperature is expected to be in the summer months, with a projected increase of 4.1 to 5.5°F and a likely range of 3.0 to 6.9°F (USGCRP 2009). This increase in temperatures is likely to cause more precipitation to fall as rain rather than snow, particularly at lower altitudes (USGCRP 2009). Heat events are projected to increase in the Pacific Northwest, particularly in south-central Washington and the western lowlands. By *mid-century* south-central Washington could experience an additional one to three heat waves annually, with other locations experiencing up to one additional heat wave each year under a moderate (A1B) emission scenario (Salathe et al. 2009).[62] The frequency of warm nights[63] is also expected to increase by roughly 7 to 20% across Washington (Salathe et al. 2009).

#### 3.8.1.3 End-of-century (2070-2100)

By the *end-of-century*, annual mean temperature is projected to have increased by 5.1 to 8.3°F, with a likely range of 3.7 to 10.0°F (USGCRP 2009). Mote and Salathe (2009)

---

[61] This information was provided by personal communication with Jay H. Lawrimore of the National Climatic Data Center.
[62] This study define heat waves as three or more days where daily heat index exceeds 90°F. Two climate models (CCSM3 and ECHAM5) are used to force the Weather Research and Forecasting Regional Model.
[63] Warm nights are defined as nights with a minimum temperature above the 90$^{th}$ percentile.

projected a smaller annual mean temperature increase of 4.8 to 6.8°F based on a low (B1) emission scenarios and a moderate (A1B) emission scenario.[64] The projections for winter and fall months are again similar to the annual mean. The mean temperature for the spring months is projected to increase by just 4.4 to 6.6°F, with a likely range of 2.5 to 8.9°F (USGCRP 2009). The summer months are projected to experience the greatest warming at 5.8 to 10.5°F, with a likely range of 4.2 to 13.1°F (USGCRP 2009).

| Pacific Northwest (Δ Temperature) | | Near-term (°F) | Mid-century (°F) | End-of-century (°F) |
|---|---|---|---|---|
| Annual | Mean | 2.2 | 3.6 – 4.3 | 5.1 – 8.3 |
| | Likely | 1.4 – 2.9 | 2.6 – 5.4 | 3.7 – 10.0 |
| | Very Likely | 0.7 – 3.7 | 1.6 – 6.4 | 2.3 – 11.8 |
| Winter | Mean | 2.1 – 2.2 | 3.5 – 3.9 | 5.1 – 7.6 |
| | Likely | 1.4 – 3.0 | 2.3 – 5.2 | 3.5 – 9.5 |
| | Very Likely | 0.6 – 3.8 | 1.1 – 6.5 | 1.8 – 11.4 |
| Spring | Mean | 1.7 – 1.9 | 3.1 – 3.4 | 4.4 – 6.6 |
| | Likely | 0.8 – 3.0 | 1.7 – 4.7 | 2.5 – 8.9 |
| | Very Likely | (0.2) – 4.1 | 0.3 – 6.1 | 0.6 – 11.2 |
| Summer | Mean | 2.5 – 2.8 | 4.1 – 5.5 | 5.8 – 10.5 |
| | Likely | 1.6 – 3.7 | 3.0 – 6.9 | 4.2 – 13.1 |
| | Very Likely | 0.7 – 4.6 | 1.8 – 8.4 | 2.5 – 15.7 |
| Fall | Mean | 2.0 – 2.2 | 3.4 – 4.2 | 4.8 – 8.4 |
| | Likely | 1.4 – 3.0 | 2.5 – 5.3 | 3.5 – 10.2 |
| | Very Likely | 0.7 – 3.6 | 1.5 – 6.4 | 2.1 – 11.9 |

Table 3-14: Annual and seasonal temperature changes for the Pacific Northwest region over the *near-term* (2010-2029), *mid-century* (2040-2059) and *end-of-century* (2080-2098) relative to 1961-1979. The range values are from low (B1) and high (A2) emissions scenarios. Data are from the USGCRP (2009).

### 3.8.2 Precipitation and Storm Events

#### 3.8.2.1 Near-term (2010-2040)

Observations across the Pacific Northwest region averaged over 1993 to 2008 suggest that annual mean precipitation has increased by less than 1% relative to the 1961-1979 time period (USGCRP 2009).[65] Mean precipitation in the Pacific Northwest is generally projected to increase in the winter, spring, and fall, while summer precipitation is projected to decrease. There is considerable disagreement across various climate models on the magnitude and direction of changes in precipitation.

---

[64] Nineteen climate models were statistically downscaled for this study, with the 5th to 95th percentile range being 4 to 9.7 degrees Fahrenheit. The reference period is 1970-1999.
[65] This information was provided by personal communication with Jay Lawrimore of the National Climatic Data Center. Annual precipitation provides some indication of regional change; however, it is not an adequate indicator when determining impacts on transportation as it masks much of the seasonal variability.

Over the next two decades, mean precipitation is projected to increase by roughly 3 to 5% for winter and fall seasons, with a likely range of approximately -3 to +12% (USGCRP 2009). The spring seasons are estimated to experience a slightly lower increase of 3%, with a likely range of -1 to +7% (USGCRP 2009). Precipitation in the summer months, on the other hand, is projected to decrease by more than 6% with a likely range of -17 to +3% (USGCRP 2009).

### 3.8.2.2 Mid-century (2040-2070)

By *mid-century*, winter mean precipitation is projected to increase by 5 to 7%, with a likely range of -3 to +17% (USGCRP 2009). Spring and fall are also projected to undergo precipitation increases of 3 to 5% with a likely range of -3 to +10%; and 5% with a likely range of -3 to +13%, respectively (USGCRP 2009). Summer, on the other hand, is projected to undergo decreases in precipitation of 8 to 17%, with a likely range of -28 to +1% (USGCRP 2009). Springtime snowpack is projected to decrease by *mid-century* in response to the increased wintertime temperature and greater occurrence of rain (versus snow). Higher average temperatures in the fall and winter will cause more precipitation to fall as rain rather than snow, and the snowpack will begin to melt earlier in the season. By the 2040s, April 1$^{st}$ snowpack is projected to decline by as much as 40% in the Cascade mountains (Payne et al. 2004, as cited in USGCRP 2009), and Leung et al. (2004) projected a 10 to 20% decline in snowfall over fall, winter, and spring in the Columbia River Basin by *mid-century* under a business as usual scenario. In addition, warm-season runoff is projected to decrease by 30% or more on the western slopes of the Cascade Mountains and by 10% in the Rocky Mountains (USGCRP 2009). The effect of the reduction in summer precipitation will be magnified by the increased evaporation as summer temperatures warm.

| Pacific Northwest (Δ Precipitation) | | Near-term (%) | Mid-century (%) | End-of-century (%) |
|---|---|---|---|---|
| **Winter** | Mean | 3 – 5 | 5 – 7 | 8 – 15 |
| | Likely | (3) – 12 | (3) – 17 | (1) – 29 |
| | Very Likely | (11) – 20 | (12) – 27 | (14) – 43 |
| **Spring** | Mean | 3 | 3 – 5 | 5 – 7 |
| | Likely | (1) – 7 | (3) – 10 | (2) – 15 |
| | Very Likely | (6) – 11 | (8) – 15 | (10) – 23 |
| **Summer** | Mean | (7) – (6) | (17) – (8) | (22) – (11) |
| | Likely | (17) – 3 | (28) – 1 | (42) – (1) |
| | Very Likely | (27) – 12 | (40) – 10 | (62) – 18 |
| **Fall** | Mean | 4 | 5 | 7 – 9 |
| | Likely | (3) – 11 | (3) – 13 | (7) – 24 |
| | Very Likely | (10) – 18 | (11) – 21 | (22) – 39 |

Table 3-15: Seasonal precipitation percent changes for the Pacific Northwest region over the *near-term* (2010-2029), *mid-century* (2040-2059) and *end-of-century* (2080-2098) relative to 1961-1979. The range values are from low (B1) and high (A2) emissions scenarios. Values in parentheses are negative values and represent decreases in precipitation. Data are from the USGCRP (2009).

By *mid-century*, Salathe et al. (2009) projected that precipitation intensity, defined in this study as annual total precipitation divided by the number of "wet" days where precipitation exceeds 1 millimeter, would increase slightly across much of Washington, with substantial increases occurring only in the northwest portion of the state under a moderate (A1B) emission scenario. The fraction of precipitation that falls on days where precipitation exceeds the 95$^{th}$ percentile was projected to increase in the eastern and western parts of the state, with decreases in the central portion along the leeward side of the Cascade Mountains (Salathe et al. 2009).

### 3.8.2.3 End-of-century (2070-2100)

By the *end-of-century*, seasonal mean precipitation is projected to have increased substantially in the winter months by 8 to 15%, with a likely range of -1 to +29% (USGCRP 2009). Spring and fall seasonal mean precipitation totals are projected to increase by 5 to 7% with a likely range of -2 to +15%; and 7 to 9% with a likely range of -7 to +24%, respectively (USGCRP 2009). The summer season is projected to continue to experience substantial declines in precipitation of 11 to 22%, with a likely range of +1 to +42% (USGCRP 2009). Diffenbaugh (2005) projected an increase of up to 10 extreme precipitation events per year in the Pacific Northwest (up to a 140% increase) under a higher (A2) emission scenario with some variation depending on location within the region.[66]

### 3.8.3 Sea-Level Rise

Global sea-level rise (SLR) of 7 to 79" (18cm to 2.0m) is projected for 2100 (see section 3.1.3 for discussion on global and local sea-level rise). SLR at the local/regional level is influenced by multiple factors, including sedimentation and erosion, ocean circulation, gravitationally induced changes, ocean density (affected by regional changes in ocean salinity and ocean temperature), and vertical motion of the land (subsidence or uplift). In the 20$^{th}$ century, the relative sea-level rise for the Pacific Northwest exhibits large variability, with locations across the region exhibiting both greater and lesser rise than the global trend.[67] For the Northwest region, sea-level rise is compounded by increased beach erosion and increased winter rainfall, which could saturate soils in the coastal bluffs leading to landslides (USGCRP 2009).

The following discussion describes studies providing local sea-level rise projections for this region. As noted in section 3.1.3 above, making local or regional projections is highly uncertain, given the incomplete understanding of some of the effects that can take

---

[66] This study defines extreme precipitation events as the number of days where precipitation exceeds the 95$^{th}$ percentile. A regional model (RegCM3) was used and projections for 2071-2095 are compared against a 1961 to 1985 baseline.
[67] The National Water Level Observation provides historic sea level trends for approximately 11 stations for the Pacific Northwest U.S. coastline. These measurements are provided by NOAA and include stations that provide sea-level trends over a 30 year span or, in most cases, much longer (NOAA 2010). The Pacific Northwest trend is compared against the global observed trend of sea-level rise of $1.8 \pm 0.5$ mm yr$^{-1}$ from 1961 to 2003 (IPCC 2007a).

place at the local level. There is no study that considers all these factors, nor is there a consistent methodology applied across these studies. Therefore, local SLR projections, while informative, should be considered carefully, and with a clear understanding of what factors each study includes or excludes.

Mote et al. (2008) estimated local sea-level rise at three different locations along the Washington state coast: Northwest Olympic Peninsula, the central and southern coast, and Puget Sound. For advisory purposes only, this study provides projections of local sea-level rise at these locations, allowing for changes in: (i) global sea-level rise, (ii) coastal elevation from the vertical movement of land at the different locations, and (iii) local wind patterns that push water toward or away from the coast. Mid-century and end-of-century estimates of the impact of vertical land motion on sea-level rise are provided for each location, with the end-of-century estimate of local vertical uplift being 15.7" (40 cm) for the Northeast Olympic Peninsula, 3.9" (10 cm) for the Central and the Southern Coast, and no change for Puget Sound. Assuming these rates remain constant in the future, studies that project global average sea-level rise should be lessened by these amounts to obtain estimates of local sea-level change.

Sea level along the Washington coastline can undergo considerable seasonal variability, with mean sea levels being 20" (50 cm) higher during the winter months compared with the summer months. The seasonal variability is explained by shifts in atmospheric circulation (winds) directly affecting the ocean elevation; that is, a northward wind can push water towards the shore increasing ocean elevation. Based on the same premise, an El Niño event can further increase sea level by 12" (30 cm). Based on an analysis of 30 scenarios, this study finds the changes in projected wintertime northward wind to range from minimal (suggesting a less than 1" reduction in mean sea level for 2050 and 2100) to increased strength (suggesting as much as a 6" (15 cm) increase in mean sea level for 2050-2099 compared with 1950-1999).[68]

---

[68]Local sea-level rise at three locations along the Northwest coast mid-century (i.e., 2050) and at end of century (i.e., 2100), projected using estimates of regional vertical land motion, atmospheric dynamics, and global sea-level rise projected by IPCC (2007a) under B1 (low emissions) and A1F1 (high emissions) scenarios. Sea-level rise is given relative to 1980-1999 mean sea level. The study caveats the results to be used for advisory purposes only. See Climate Change Effects Typology Matrix for projected results. (Mote et al. 2008)

## 3.9 Alaska

### 3.9.1 Temperature

#### 3.9.1.1 Near-term (2010-2040)

Within the next several decades, Alaska may experience an increase in annual average temperature of about 2.4 to 2.6°F, with a likely range of 1.5 to 3.6°F (USGCRP 2009). Fall and spring seasonal temperature projections are similar to the annual average. The greatest warming is projected to occur in the winter, when temperatures are projected to increase 3.1 to 4.0°F and a likely range of 1.0 to 5.9°F (USGCRP 2009). Summers are projected to warm on average by 1.3°F, with a likely range of 0.6 to 2.0°F, although summer is likely to be the least affected of all seasons (USGCRP 2009). The CCSP (2008) projects a slightly higher average annual temperature increase of 3.6°F in Alaska by 2030. It is interesting to note that while Alaska has warmed at more than twice the rate of the national average (USGCRP 2009), the magnitude of the projections for the annual mean average in Alaska is consistent with the projected annual mean averages of other U.S. regions. The seasonal averages are also consistent with other northern regions, with the winter months demonstrating the greatest potential increase. The higher temperatures are reducing sea ice and glacier mass or glacier extent, leading to an earlier spring snowmelt, and affecting permafrost (USGCRP 2009). Permafrost warming leads to land subsidence, directly affecting highway systems (Larsen et al. 2008). These higher temperatures will also lead to greater evaporation rates, potentially reducing soil moisture and leading to reductions in the water levels in closed-basin lakes (USGCRP 2009).

#### 3.9.1.2 Mid-century (2040-2070)

By *mid-century*, the average projected increase in annual average temperature is 4.3°F with a likely range of 3.6 to 5.0°F (USGCRP 2009). Fall and spring are expected to continue to experience similar seasonal mean temperature changes compared with the annual average, though with some differences. Seasonal mean temperatures for fall are projected to increase slightly more than the annual average, by 4.4 to 4.9°F, with a likely range of 3.5 to 5.6°F (USGCRP 2009). Mean spring temperature is projected to increase by 3.9 to 4.0°F, with a likely range of 3.6 to 11.5°F—slightly less than the annual average. As with the earlier projected period, the largest increase in average temperatures occurs during the winter months, with the seasonal mean temperature increase of 6.2 to 6.4°F and a likely range of 5.1 to 7.7°F (USGCRP 2009). Summer is expected to continue to be the least affected season, with a projected increase of 2.1 to 2.5°F and a likely range of 0.7 to 3.5°F (USGCRP 2009).

| Alaska (Δ Temperature) | | Near-term (°F) | Mid-century (°F) | End-of-century (°F) |
|---|---|---|---|---|
| Annual | Mean | 2.4 – 2.6 | 4.3 | 6.7 – 9.9 |
|  | Likely | 1.5 – 3.6 | 3.6 – 5.0 | 4.6 – 11.7 |
|  | Very Likely | 0.4 – 4.7 | 2.9 – 5.7 | 2.4 – 13.5 |
| Winter | Mean | 3.1 – 4.0 | 6.2 – 6.4 | 9.9 – 14.5 |
|  | Likely | 1.0 – 5.9 | 5.1 – 7.7 | 7.5 – 17.4 |
|  | Very Likely | (1.1) – 7.9 | 3.8 – 9.0 | 5.1 – 20.2 |
| Spring | Mean | 2.3 – 2.6 | 3.9 – 4.0 | 6.2 – 9.1 |
|  | Likely | 0.6 – 4.7 | 2.5 – 5.5 | 3.6 – 11.5 |
|  | Very Likely | (1.5) – 6.8 | 1.1 – 7.0 | (0.9) – 14.0 |
| Summer | Mean | 1.3 | 2.1 – 2.5 | 3.9 – 5.9 |
|  | Likely | 0.6 – 2.0 | 0.7 – 3.5 | 1.5 – 8.7 |
|  | Very Likely | (0.1) – 2.8 | (0.8) – 5.0 | (1.0) – 13.9 |
| Fall | Mean | 2.3 – 2.8 | 4.4 – 4.9 | 7.0 – 10.0 |
|  | Likely | 1.6 – 3.1 | 3.5 – 5.6 | 4.9 – 11.5 |
|  | Very Likely | 0.8 – 3.7 | 2.6 – 6.3 | 2.8 – 13.0 |

Table 3-16: Annual and seasonal temperature changes for Alaska over the *near-term* (2010-2029), *mid-century* (2040-2059) and *end-of-century* (2080-2098) relative to 1961-1979. The range values are from low (B1) and high (A2) emissions scenarios. Parentheses represent negative projections. Data are from the USGCRP (2009) and are based on the five climate models identified as the top performers for Alaska by Walsh et al. (2008). The results listed in the typology matrix for Alaska are provided for the suite of models consistent with the other U.S. regions.

### 3.9.1.3 End-of-century (2070-2100)

By late century, Alaska will likely see an increase of 6.7 to 9.9°F in average annual temperature, with a likely range of 4.6 to 11.7°F (USGCRP 2009). Spring and fall seasonal mean temperature projections continue to follow the annual trends. Winter will continue to be the most affected of the seasons, with an average seasonal temperature increase of 9.9 to 14.5°F and a likely range of 7.5 to 17.4°F (USGCRP 2009). Summer is likely to remain the least affected of the seasons, with projected increases of 3.9°F to 5.9°F and a likely range of 1.5 to 8.7°F (USGCRP 2009). By the end of the 21st century, northern Alaska could see surface temperatures increase by more than 9.0°F (5°C) (IPCC 2007a) and permafrost temperatures on the Seward Peninsula could increase by 0 to 5.8°F (Busey et al. 2008) under a moderate (A1B) emission scenario.[69] Additionally, by the 2080 to 2100 period, an extreme heat event that currently has a 5% chance of occurring per year is likely to have a 10% chance of occurring within a given year under a higher (A2) emission scenario (USGCRP 2009).

---

[69] This study uses the TOPP numerical model and compares 2090-2100 projections relative to 2001-2004 temperatures. Largest change is on the coast with some high elevation areas becoming slightly colder.

### 3.9.2 Precipitation and Storm Events

#### 3.9.2.1 *Near-term (2010-2040)*

The winter months are projected to have the most noticeable change in seasonal precipitation. Within the next several decades, Alaska is likely to experience an increase in winter mean precipitation of 6 to 9% with a likely range of +1 to +16% (USGCRP 2009). Similar to projected temperatures, summer will likely be the least affected of the seasons in terms of precipitation. Summer is expected to experience an increase in mean precipitation of approximately 6% with a likely range of +3 to +9% (USGCRP 2009). Spring and fall are estimated to receive additional precipitation of 5 to 8% with a likely range of +0 to +15%; and 7 to 8% with a likely range of +4 to +11%, respectively (USGCRP 2009).

#### 3.9.2.2 *Mid-century (2040-2070)*

By *mid-century*, winters in Alaska may continue to experience the largest seasonal increase in mean precipitation, 15 to 17%, with a likely range of 10 to 20% (USGCRP 2009). Spring is likely to be the least affected of the seasons, with changes in mean precipitation estimated to by an increase of 9 to 12% with a likely range of +3 to +15%. The average projected increases for fall and summer mean precipitation are moderate compared with the other seasons at 10 to 14%, with a likely range of +8 to +16%; and 11 to 13%, with a likely range of +7 to +17%, respectively (USGCRP 2009).

| Alaska (Δ Precipitation) | | Near-term (%) | Mid-century (%) | End-of-century (%) |
|---|---|---|---|---|
| **Winter** | Mean | 6 – 9 | 15 – 17 | 23 – 37 |
| | Likely | 1 – 16 | 10 – 20 | 18 – 48 |
| | Very Likely | (5) – 23 | 4 – 25 | 13 – 59 |
| **Spring** | Mean | 5 – 8 | 9 – 12 | 18 – 30 |
| | Likely | 0 – 15 | 3 – 15 | 12 – 34 |
| | Very Likely | (7) – 22 | (3) – 21 | 7 – 38 |
| **Summer** | Mean | 6 | 11 – 13 | 17 – 23 |
| | Likely | 3 – 9 | 7 – 17 | 14 – 29 |
| | Very Likely | (1) – 13 | 3 – 20 | 11 – 36 |
| **Fall** | Mean | 7 – 8 | 10 – 14 | 17 – 30 |
| | Likely | 4 – 11 | 8 – 16 | 9 – 39 |
| | Very Likely | 1 – 13 | 6 – 19 | 1 – 46 |

Table 3-17: Seasonal precipitation percent changes for Alaska over the *near-term* (2010-2029), *mid-century* (2040-2059) and *end-of-century* (2080-2098) relative to 1961-1979. The range values are from low (B1) and high (A2) emissions scenarios. Values in parentheses are negative values and represent decreases in precipitation. Data are from the USGCRP (2009) and are based on the five climate models identified as the top performers for Alaska by Walsh et al. (2008). The results listed in the typology matrix for Alaska are provided for the suite of models consistent with the other U.S. regions.

### 3.9.2.3 End-of-century (2070-2100)

By the *end-of-century*, winter will likely continue to see the greatest increases in total seasonal precipitation. Winters in Alaska could see an 23 to 37% increase in mean precipitation, with a likely range of +18 to +48% (USGCRP 2009). The fall season is likely to experience the smallest increases of 17 to 30%, with a likely range of +9 to +39% (USGCRP 2009). Spring and summer continue to demonstrate moderate increases compared with the other seasons, at 18 to 30%, with a likely range of +12 to +34%; and 17 to 23%, with a likely range of +14 to +29%, respectively (USGCRP 2009).

Storm activity may increase as the Pacific storm track moves northward and sea surface temperatures increase (USGCRP 2009). These storms are expected to be situated over oceans with less ice cover (both in magnitude and seasonal duration), and may increase in frequency and intensity as the warmer ocean may supply these storm with more heat and moisture (USGCRP 2009).

### 3.9.3 Sea-Level Rise

Global sea-level rise (SLR) of 7 to 79" (18cm to 2.0m) is projected for 2100 (see section 3.1.3 for discussion on global and local sea-level rise). SLR at the local/regional level is influenced by multiple factors, including sedimentation and erosion, ocean circulation, gravitationally induced changes, ocean density (affected by regional changes in ocean salinity, and ocean temperature), and vertical motion of the land (subsidence or uplift). In the $20^{th}$ century, the relative sea-level rise for Alaska exhibits large variability, with some locations demonstrating slightly greater rise compared to observed global sea-level rise, and with other stations demonstrating considerably lower.[70] Parts of Alaska are undergoing uplift in response to glacial ice loss and active regional tectonic deformation at a rate considered to keep pace with rising global sea levels. In fact, relative sea level in the southeast and south central Gulf of Alaska coastal area may actually decrease substantially (Larsen et al. 2004; Kelly et al. 2007). It should be noted that under high sea-level rise projections, it is possible that sea-level rise rates may approach or exceed uplift rates.

---

[70] The National Water Level Observation provides historic sea level trends for approximately 15 stations for the Alaska U.S. coastline. These measurements are provided by NOAA and include stations that provide sea level trends over a 30-year span or, in most cases, much longer (NOAA 2010). This trend is compared against the global observed trend of sea-level rise of $1.8 \pm 0.5$ mm/yr from 1961 to 2003 (IPCC 2007a).

## 3.10 Hawaii

### 3.10.1 Temperature

Over the next few decades, annual mean temperatures in Hawaii are projected to increase by about 1.8°F, with a likely range of 1.0 to 2.5°F (USGCRP 2009). The seasonal mean temperatures and ranges are very similar to the annual mean, with an even consistency of warming throughout the year. This relationship continues through to the *end-of-century*. By *mid-century*, the annual mean temperatures may increase by 2.7 to 3.3°F with a likely range of 2.0 to 4.0°F (USGCRP 2009). By the *end-of-century*, the annual mean temperature is projected to have increased by 3.9 to 6.7°F with a likely range of 2.8 to 7.8°F (USGCRP 2009). Some small seasonal variations from the annual mean are projected for the *end-of-century*.

| Hawaii (Δ Temperature) | | Near-term (°F) | Mid-century (°F) | End-of-century (°F) |
|---|---|---|---|---|
| Annual | Mean | 1.7 – 1.8 | 2.7 – 3.3 | 3.9 – 6.7 |
|  | Likely | 1.0 – 2.5 | 2.0 – 4.0 | 2.8 – 7.8 |
|  | Very Likely | 0.3 – 3.2 | 1.2 – 4.6 | 1.8 – 8.9 |
| Winter | Mean | 1.7 | 2.7 – 3.2 | 3.8 – 6.4 |
|  | Likely | 1.0 – 2.4 | 1.9 – 3.9 | 2.7 – 7.5 |
|  | Very Likely | 0.3 – 3.0 | 1.1 – 4.6 | 1.6 – 8.5 |
| Spring | Mean | 1.6 – 1.8 | 2.7 – 3.2 | 3.8 – 6.3 |
|  | Likely | 0.9 – 2.6 | 1.9 – 3.7 | 2.8 – 7.2 |
|  | Very Likely | 0.1 – 3.4 | 1.1 – 4.3 | 1.8 – 8.1 |
| Summer | Mean | 1.8 | 2.7 – 3.4 | 3.9 – 6.7 |
|  | Likely | 1.0 – 2.6 | 1.9 – 4.0 | 2.8 – 7.9 |
|  | Very Likely | 0.1 – 3.4 | 1.1 – 4.7 | 1.7 – 9.2 |
| Fall | Mean | 1.8 | 2.8 – 3.5 | 4.0 – 7.2 |
|  | Likely | 1.0 – 2.6 | 2.0 – 4.4 | 2.8 – 8.7 |
|  | Very Likely | 0.3 – 3.3 | 1.2 – 5.2 | 1.6 – 10.3 |

Table 3-18: Annual and seasonal temperature changes for Hawaii over the *near-term* (2010-2029), *mid-century* (2040-2059) and *end-of-century* (2080-2098) relative to 1961-1979. The range values are from low (B1) and high (A2) emissions scenarios. Data are from the USGCRP (2009).

### 3.10.2 Precipitation and Storm Events

#### 3.10.2.1 *Near-term* (2010-2040)

While the temperature is not expected to have much seasonal variation, precipitation is projected to vary seasonally. Over the next few decades, the greatest decrease in precipitation is projected to occur during the summer months, with declines of 3 to 5% and a likely range of -18 to +8% (USGCRP 2009). Winter mean precipitation is also projected to decrease by approximately 2% and a likely range of -15 to +10% (USGCRP

2009). Fall mean precipitation is projected to change between -1 to +2%, with a likely range of -11 to +11% (USGCRP 2009). Spring mean precipitation is projected to change by 0 to +1%, with a likely range of -10 to +11% (USGCRP 2009).

| Hawaii (Δ Precipitation) | | Near-term (%) | Mid-century (%) | End-of-century (%) |
|---|---|---|---|---|
| **Winter** | Mean | (2) | (3) – (2) | (4) – (1) |
| | Likely | (15) – 10 | (21) – 18 | (25) – 22 |
| | Very Likely | (27) – 22 | (40) – 37 | (47) – 46 |
| **Spring** | Mean | 0 – 1 | (2) | (5) – 6 |
| | Likely | (10) – 11 | (13) – 9 | (20) – 9 |
| | Very Likely | (20) – 21 | (24) – 20 | (34) – 22 |
| **Summer** | Mean | (5) – (3) | (3) –(1) | (1) – 5 |
| | Likely | (18) – 8 | (23) – 21 | (42) – 51 |
| | Very Likely | (30) – 20 | (45) – 43 | (88) – 98 |
| **Fall** | Mean | (1) – 2 | 3 – 6 | 1 – 19 |
| | Likely | (11) – 11 | (15) – 27 | (38) – 75 |
| | Very Likely | (21) – 21 | (36) – 47 | (95) – 132 |

Table 3-19: Seasonal precipitation percent changes for the Hawaii over the *near-term* (2010-2029), *mid-century* (2040-2059) and *end-of-century* (2080-2098) relative to 1961-1979. The range values are from low (B1) and high (A2) emissions scenarios. Values in parentheses are negative values and represent decreases in precipitation. Data are from the USGCRP (2009).

3.10.2.2    *Mid-century* (2040-2070)

By *mid-century*, only the fall months are projected to increase in precipitation, by 3 to 6% and a likely range of -15 to +27% (USGCRP 2009). The other three seasons are projected to decrease in precipitation. The winter and spring months are similar in projected mean change but not when comparing the likely range. The winter months continue to project a decrease in mean precipitation by 2 to 3% with a likely range of -21 to +18% (USGCRP 2009). The spring months may have a decrease in mean precipitation by 2%, with a likely range of -13 to +9% (USGCRP 2009). The summer months have a larger variability of decreased mean precipitation of 1 to 3%, with a likely range of -23 to +21% (USGCRP 2009).

3.10.2.3    *End-of-century* (2070-2100)

By the *end-of-century*, the winter months continue to demonstrate a decrease in mean precipitation of 1 to 4%, with a likely range of -25 to +22% (USGCRP 2009). The spring and summer months have a mean change in precipitation that crosses direction. The changes in spring mean precipitation vary between a decrease of 5% to an increase of 6%, with a likely range of -20 to +9% (USGCRP 2009). The changes in summer mean precipitation vary between a decrease of 1% to an increase of 5%, with a large likely range of -42 to +51% (USGCRP 2009). The fall months exhibit an increase in the mean change for precipitation, of 1 to 19%, with a likely range of -38 to +75% (USGCRP 2009).

### 3.10.3 Sea-Level Rise

Global sea-level rise (SLR) of 7 to 79" (18cm to 2.0m) is projected for 2100 (see section 3.1.3 for discussion on global and local sea-level rise). SLR at the local/regional level is influenced by multiple factors, including sedimentation and erosion, ocean circulation, gravitationally induced changes, ocean density (affected by regional changes in ocean salinity and ocean temperature), and vertical motion of the land (subsidence or uplift). In the 20$^{th}$ century, the relative sea-level rise for Hawaii is generally similar to that of observed global sea-level rise.[71] Some islands are particularly vulnerable to sea-level rise and storm surge, with resulting shoreline erosion. For example, the Northwestern Hawaiian Islands are low-lying islands particularly vulnerable to sea-level rise (USGCRP 2009).

---

[71] The National Water Level Observation provides historical sea-level trends for 5 stations in Hawaii. These measurements are provided by NOAA and include stations that provide sea-level trends over a 30-year span or, in most cases, much longer (NOAA 2010). This trend is compared against the global observed trend of sea-level rise of 1.8 ± 0.5 mm/yr from 1961 to 2003 (IPCC 2007a).

## 3.11 Puerto Rico

The information provided below represents the Caribbean region.

### 3.11.1 Temperature

Over the next few decades, the annual mean temperatures are projected to increase by about 1.7°F, with a likely range of 1.2 to 2.1°F (USGCRP 2009). The magnitude of the warming is expected to be roughly similar in all four seasons through to the *end-of-century*. By *mid-century*, the annual mean temperatures may increase by 2.5 to 3.1°F, with a likely range of 2.0 to 3.5°F (USGCRP 2009). By the *end-of-century*, the annual mean temperature is projected to have increased by 3.6 to 6.1°F, with a likely range of 2.7 to 6.8°F (USGCRP 2009).

| Caribbean (Δ Temperature) | | Near-term (°F) | Mid-century (°F) | End-of-century (°F) |
|---|---|---|---|---|
| Annual | Mean | 1.6 – 1.7 | 2.5 – 3.1 | 3.6 – 6.1 |
| | Likely | 1.2 – 2.1 | 2.0 – 3.5 | 2.7 – 6.8 |
| | Very Likely | 0.8 – 2.5 | 1.4 – 3.9 | 1.9 – 7.5 |
| Winter | Mean | 1.6 | 2.4 – 3.0 | 3.5 – 5.8 |
| | Likely | 1.1 – 2.0 | 1.9 – 3.5 | 2.6 – 6.6 |
| | Very Likely | 0.8 – 2.5 | 1.3 – 3.9 | 1.7 – 7.5 |
| Spring | Mean | 1.5 – 1.6 | 2.5 – 3.0 | 3.5 – 5.8 |
| | Likely | 1.1 – 2.0 | 1.9 – 3.4 | 2.7 – 6.5 |
| | Very Likely | 0.6 – 2.5 | 1.3 – 3.7 | 1.8 – 7.2 |
| Summer | Mean | 1.7 – 1.8 | 2.6 – 3.2 | 3.7 – 6.2 |
| | Likely | 1.2 – 2.1 | 2.1 – 3.6 | 2.8 – 6.9 |
| | Very Likely | 0.8 – 2.6 | 1.5 – 4.0 | 1.9 – 7.6 |
| Fall | Mean | 1.7 – 1.8 | 2.7 – 3.3 | 3.7 – 6.4 |
| | Likely | 1.3 – 2.1 | 2.1 – 3.7 | 2.8 – 7.1 |
| | Very Likely | 0.9 – 2.5 | 1.5 – 4.1 | 1.9 – 7.9 |

Table 3-20: Annual and seasonal temperature changes for the Caribbean over the *near-term* (2010-2029), *mid-century* (2040-2059) and *end-of-century* (2080-2098) relative to 1961-1979. The range values are from low (B1) and high (A2) emissions scenarios. Data are from the USGCRP (2009).

### 3.11.2 Precipitation and Storm Events

Overall, the Caribbean is expected to experience significant reductions in precipitation for almost all seasons and time horizons. The greatest reduction of precipitation is projected for the summer and spring months, and the least reduction in the winter and fall months.

### 3.11.2.1   Near-term (2010-2040)

Over the next few decades, the largest reduction in mean precipitation is expected in the summer months of 10 to 7%, with a likely range of -16 to +1% (USGCRP 2009). Winter mean precipitation is also projected to decrease by approximately 1 to 3% and a likely range of -9 to +5% (USGCRP 2009). Fall mean precipitation is projected to decrease by 1 to 2%, with a likely range of -9 to +6% (USGCRP 2009). Spring mean precipitation is projected to decrease by 6 to 7%, with a likely range of –15 to +2% (USGCRP 2009).

| Caribbean (Δ Precipitation) | | Near-term (%) | Mid-Term (%) | Long-Term (%) |
|---|---|---|---|---|
| **Winter** | Mean | (3) – (1) | (5) – (3) | (8) – (2) |
| | Likely | (9) – 5 | (14) – 5 | (22) – 6 |
| | Very Likely | (15) – 11 | (22) – 12 | (35) – 19 |
| **Spring** | Mean | (7) – (6) | (16) – (8) | (28) – (9) |
| | Likely | (15) – 2 | (25) – 0 | (39) – 2 |
| | Very Likely | (24) – 11 | (33) – 9 | (51) – 13 |
| **Summer** | Mean | (10) – (7) | (18) – (12) | (36) – (14) |
| | Likely | (16) – 1 | (31) – (1) | (52) – 0 |
| | Very Likely | (23) – 8 | (44) – 11 | (68) – 14 |
| **Fall** | Mean | (2) – (1) | (4) – (3) | (9) – (4) |
| | Likely | (9) – 6 | (15) – 7 | (28) – 10 |
| | Very Likely | (16) – 12 | (26) – 18 | (47) – 29 |

Table 3-21: Seasonal precipitation percent changes for the Caribbean over the *near-term* (2010-2029), *mid-century* (2040-2059) and *end-of-century* (2080-2098) relative to 1961-1979. The range values are from low (B1) and high (A2) emissions scenarios. Values in parentheses are negative values and represent decreases in precipitation. Data are from the USGCRP (2009).

### 3.11.2.2   Mid-century (2040-2070)

*Mid-century* precipitation is expected to continue to have declined in all seasons. The winter and fall months are projected to have similar mean changes; but their likely ranges are noticeably different. The fall months are projected to decrease in precipitation by 3 to 4%, with a likely range of -15 to +7% (USGCRP 2009). The winter months are projected to decrease in mean precipitation by 3 to 5%, with a likely range of -14 to +5% (USGCRP 2009). The spring months are projected to have a substantial decrease in mean precipitation by 8 to16%, with a likely range of -25 to 0% (USGCRP 2009). The summer months are expected to have a larger decrease in mean precipitation of 12 to 18%, with a likely range of -31 to -1% (USGCRP 2009).

### 3.11.2.3  End-of-century (2070-2100)

Precipitation totals are expected to continue to decline through the *end-of-century*. The winter months are projected to continue to experience a decrease in mean precipitation of 2 to 8%, with a likely range of -22 to +6% (USGCRP 2009). The spring mean precipitation is projected to decrease by 9 to 28%, with a likely range of -39 to +2% (USGCRP 2009). The summer mean precipitation is projected to decrease by 14 to 36%, with a large likely range of -52 to 0% (USGCRP 2009). The fall months are projected to experience a decrease in mean precipitation of 4 to 9%, with a likely range of -28 to +10% (USGCRP 2009).

### 3.11.3 Sea-level rise

Global sea-level rise (SLR) of 7 to 79" (18cm to 2.0m) is projected for 2100 (see section 3.1.3 for discussion on global and local sea-level rise). SLR at the local/regional level is influenced by multiple factors, including sedimentation and erosion, ocean circulation, gravitationally induced changes, ocean density (affected by regional changes in ocean salinity and ocean temperature), and vertical motion of the land (subsidence or uplift). In the 20$^{th}$ century, the relative sea-level rise for the Caribbean is generally slightly less than that of global sea-level rise.[72]

---

[72] The National Water Level Observation provides historical sea-level trends for 2 stations in Puerto Rico. These measurements are provided by NOAA and include stations that provide sea-level trends over a 30 year span or, in most cases, much longer (NOAA 2010). This trend is compared against the global observed trend of sea-level rise of $1.8 \pm 0.5$ mm yr$^{-1}$ from 1961 to 2003 (IPCC 2007a).

# 4 Future Work

## 4.1 Improvements to current projections

The climate projections provided in this report are based mostly on the results from a collection of global climate models aggregated to the multi-state regional scale. This set of data is uniformly available across all regions in the United States. Additional information of downscaled projections has been provided at the state or sub-state level as available. In the coming years, the data provided in this report will likely become outdated as model simulations and downscaling techniques that provide regional- and local-scale climate projections continue to evolve. Ideally, information downscaled to the local level will become uniformly available, a task perhaps for a National Climate Service.

### 4.1.1 Uncertainty

Uncertainty in the projections provided in this report includes model uncertainty and emission uncertainty (i.e., how closely emissions and concentrations of GHGs in the future match the two scenarios—the B1 and A2 scenarios—used in this report), while natural variability is not directly assessed. As climate science progresses, the degree of uncertainty is likely to be reduced—particularly for regional-scale projections. For example, Hawkins and Sutton (2009) suggest that the uncertainty associated with regional projections in the *near-term*, dominated by model uncertainty and natural variability, could be significantly reduced through scientific progress.

### 4.1.2 Precipitation, Storm Events, and Sea-level rise

Certain climate effects are particularly problematic to model, while others are simply not available at the regional or finer scale. Of the climate effects discussed by region in this report, precipitation and storm events are particularly challenging. For many regions, there is a large range of plausible values for precipitation that may swing dramatically between significant increases to significant decreases. As science hones its skills representing the water cycle and the degree of uncertainty decreases, these ranges may narrow. In addition, changes in precipitation extremes (e.g., precipitation frequency, duration, and intensity) are extremely important for planning purposes but sparsely addressed in the literature at the regional or local scale.

Continued research in building techniques for assessing projected storm growth is another area of concern for highway planners. Across the United States, storms of interest can vary between small-scale convective storms (i.e., thunderstorms) to larger-scaled cyclones (i.e., Nor'easters and other mid-latitude extratropical cyclones, tropical storms, and hurricanes). There are many factors that contribute to whether a specific storm will grow or dissipate. For example, the growth of a tropical storm into a hurricane depends on such factors as wind shear, vertical temperature structure over the ocean, and sea surface temperatures, all of which can vary differently in a changing climate system. It is a challenge for climate models to capture these various components and then project how, under a changing climate, they may affect storms.

The global estimate of sea-level rise associated with thermal expansion and ice melting is an area of intense research, with studies continually being published with new methods and refined estimates. It is likely that the global estimate of sea-level rise will continue to evolve in the coming years. This report also discusses projected local sea-level rise where global estimates are applied and adjusted through considering local factors such as vertical land motion, ocean circulation, erosion and sedimentation, and ocean density. As these projections can vary significantly between nearby locations due to local variability of these contributing factors, local-scale efforts that address these factors would provide additional data to assist highway planners. Currently, no study presented in this report provides all inclusive sea-level rise projections taking all of these factors into account. It is understood that scientists are improving their skills and methods in estimating the observed and projected local variability; the results of these efforts will provide highway planners with increased accuracy in estimating local sea-level rise.

## 4.2 Additional climate variables or enhanced techniques

Changes in solar radiation and humidity can affect highway systems. Though climate models do provide projections of these variables, the work is still rudimentary in using these projections for addressing climate change impacts.[73] . In the future, it is likely that data and projections for these variables will be available in a form useful to highway planners.

This report broadly discusses the impacts of the climate projections by treating each climate variable separately. New techniques may become available that account for the impact of multiple climate effects on a particular component of the highway system. For example, many in the health sector have replaced the use of maximum temperature with a heat index (a combination of maximum temperature and relative humidity) in determining the impact of heat events on mortality. Similar climate variable combinations may be determined that perform better in assessing the projected impacts on the highway system. For example, the assessments of projected impacts could be examined through engineering design specifications associated with worst-case scenarios. Instead of determining impact by isolating each climate effect (i.e., investigating the projected change associated with each engineering design specification), this effort would take the next step to combine the impacts associated with all the design specifications, thereby providing an approximation of impacts from multiple climate effects.

---

[73] Though climate models provide humidity as an output variable, much of the available literature that investigates humidity projections use a relationship based on temperature projections.

# 5 Acronyms and Glossary

**Acronyms**

**BCCR BCM2.0**: Bjerknes Centre for Climate Research (BCCR) Bergen Climate Model (BCM) Version 2, Norway

**CCSM3.0**: National Center for Atmospheric Research (NCAR) Community Climate System Model, Version 3.0

**CGCM3.1**: Canadian Centre for Climate Modelling and Analysis Coupled General Circulation Model, Version 3.1, Canada

**CNRM**: Centre National de Recherches Météorologiques, France

**CCSP**: Climate Change Science Program

**CIG**: Climate Impacts Group

**CSIRO**: Commonwealth Scientific and Industrial Research Organization, Australia

**DOT**: Department of Transportation

**FHWA**: Federal Highway Administration

**GHG**: Greenhouse gas

**GCM**: General Circulation Model

**GFDL CM 2.0**: Geophysical Fluid Dynamics Laboratory Climate Model, Version 2.0

**GFDL CM 2.1**: Geophysical Fluid Dynamics Laboratory Climate Model, Version 2.1

**GISS AOM**: Goddard Institute for Space Studies, Atmosphere-Ocean Model

**HadCM3**: Hadley Centre Coupled Model, Version 3, United Kingdom

**HadGEM1**: Hadley Centre Global Environmental Model, Version 1, United Kingdom

**IAP FGOALS-g1.0**: Institute of Atmospheric Physics (IAP), Flexible Global Ocean Atmosphere Land System (FGOALS) model, Gridpoint Version 1.0, China

**INM-CM3.0**: Institute of Numerical Mathematics Coupled Model, Version 3.0, Russia

**IPCC**: Intergovernmental Panel on Climate Change

**IPCC AR4**: Intergovernmental Panel on Climate Change Fourth Assessment Report

**IPSL CM4**: L'Institut Pierre-Simon Laplace Coupled Model, Version 4, France

**MIROC3.2(hires)**: Model for Interdisciplinary Research on Climate, High Resolution Version, Center for Climate System Research, Japan

**MIROC3.2(medres)**: Model for Interdisciplinary Research on Climate, Medium Resolution Version, Center for Climate System Research, Japan

**MIUB ECHO-G**: Meteorological Institute University of Bonn (MIUB), ECHO-G, Germany

**MPI ECHAM**: Max Planck Institute for Meteorology (MPI), ECHAM model, Version 5, Germany

**MRI CGCM 2.3.2a**: Meteorological Research Institute (MRI), Coupled General Circulation Model (CGCM), Version 2.3.2a, Japan

**NCAR**: National Center for Atmospheric Research

**NCDC**: National Climatic Data Center, National Oceanic & Atmospheric Administration (NOAA)

**NRC**: National Research Council

**PCM**: Parallel Climate Model

**SAP**: Synthesis and Assessment Product

**SRES**: Special Report Emission Scenarios

**USGCRP**: United States Global Change Research Program

**Glossary**

This information is summarized from a collection of sources (USGCRP 2009, CCSP 2008a, CCSP 2009b, IPCC 2007a, Lutgens and Tarbuck 2007, Ahrens 2007).

**1-in-20 weather event**
An extreme weather event (e.g., extreme temperature, heat wave, rainfall, storms, storm surges) of a severity that has a probability of occurring only once every 20 years (i.e., a 5% chance of occurring in any given year, or with a severity at the $95^{th}$ percentile of similar events). However, an event occurring in one year does not preclude the event from occurring the following year or even the following week.

**Annual mean temperature**
The arithmetic mean, or average, of daily temperatures for a given year.

**Climate change effects**
Changes in climate variables that are brought about by climate change.

**Climate models**
A model that incorporates the principles of physics, chemistry, and biology into a mathematical model of climate.

**Climate stressors**
For the purposes of this report, climate stressors are climate effects that affect the design, construction, operation, and/or maintenance of transportation infrastructure.

**Climate variables**
Physical characteristics of climate, such as temperature, humidity, level of precipitation, frequency of storms, sea-level rise, that can be quantified using climate models.

**Diurnal temperature range**

The variation in temperature between the minimum (lowest) temperature and the maximum (highest) temperature for a given day.

**Downscaling**
Mathematical techniques that have been developed to transform the projected climate effects of global climate models to a regional scale. Global climate models partition the world into large grids of cells that hide regional details; downscaling is used to provide climate effect results at a finer resolution. Downscaling techniques generally can be classified as either statistical downscaling or dynamic downscaling.

**Downscaling, statistical**
Statistical downscaling determines a relationship between the climate model output of a climate effect for a past 30- to 40-year time period and the observed climate effect for the same time period. This relationship is then used to downscale the projected climate model output for that particular climate effect. This approach is best when the determined relationship is robust over time; that is, the processes governing the climate effect remain fixed with time. This may not always be an appropriate method for downscaling precipitation projections (NECIA 2006).

**Downscaling, dynamical**
Dynamical downscaling uses a regional model equipped with small-scale processes and local topography. The climate model data are used as inputs around the boundaries of the regional model. This technique allows for capturing the effects of local topography. Though this process tends to be expensive and time-consuming, it does include dynamical changes in response to large-scale forcing. Given the investment for this approach, dynamical downscaled studies generally provide projections based on the results from only a single climate model.

**Effects typology matrix**
A tool to summarize projected climate effects by: geographic location, time horizon, emission scenario, and climate variable. The effects typology matrix developed in this report can be found in Appendix C.

**Emissions scenarios**
Hypothetical scenarios of future greenhouse gas emissions profiles.

**Time horizon, *End-of-century***
For the purposes of this report, *End-of-century* refers to a 2070 to 2100 time horizon.

**Evaporation**
The process through which a liquid is transformed into a gas.

**Extreme cold events**
An extreme weather event characterized by very cold temperatures. The exact definition

of what is meant by an extreme cold event can vary. For example, extreme cold events may be defined as temperatures below the 5th percentile for a given region.

**Extreme weather events**
Weather events that exhibit very severe conditions such as intense levels of precipitation or high levels of wind and can cause significant damage. These events can include intense tornado-breeding convective storms, well-developed mid-latitude cyclones (such as a nor'easter), and tropical storms (particularly ones that develop into hurricanes).

**Extreme heat day**
The maximum temperature for a given day exceeds 90°F (in some studies, this term is defined as the maximum temperature for a given day exceeds 100°F).

**Extreme heat event**
Has multiple definitions of duration and intensity, and generally is a relative measure to local climate conditions. The Centers for Disease Control defines it as a temperature 10°F or more above the average high temperature for a region, lasting several weeks. Other sources define the event by the apparent temperature (determined through a calculation based on maximum temperature and humidity) reaching above a 1-in-20-year event (or an event that has a 5% chance of occurring per year).

**Fall months**
September, October, and November (SON)

**Freeze-thaw cycle**
Period of time that elapses between freezing and thawing conditions.

**Heat waves**
Three or more days where daily heat index exceeds 90°F.

**High Temperature Day**
At or above the 95th percentile among current daily temperature records (Diffenbaugh et al 2005).

**Humidity, relative**
The ratio of air's water vapor content to its water vapor capacity.

**Hurricanes**
A tropical cyclonic storm having minimum winds of 74 miles per hour (64 knots, or 119 kilometers per hour).

**Interglacial period**
Period of warm global average temperatures between two glacial periods within an ice age. The Earth is currently in an interglacial period called the Holocene.

**Jet stream, polar**
A swift, westerly airstream in the upper troposphere that meanders in relatively narrow belts and is a result of the boundary between two surface air masses: warm southern air and cold northern air. The polar jet stream is usually located between 9 to 12 km above the Earth's surface (altitude).

**Emissions scenario; low emissions scenario (B1)**
The low emissions scenario referenced in this report corresponds with the B1 emission scenario developed by the IPCC (2000). This scenario describes a "convergent world" with low population growth and rapid changes in the global economy toward service and information sectors. It assumes reductions in material intensity, and the introduction of clean and resource-efficient technologies. (IPCC 2000)

**Time horizon, *Mid-century***
For the purposes of this report, *Mid-century* refers to a 2040 to 2070 time horizon.

**Emissions scenario; moderately high "business as usual" emission scenario (A2)**
The moderately high, or "business as usual" emission scenario referenced in this report corresponds with the A2 emission scenario developed by the IPCC (2000). "Business as usual" refers to an emission scenario that is assumed to occur without any effort to reduce greenhouse gas emissions from present practices. The A2 scenario describes a "fragmented" future with less cooperation between world governments, high population growth, and regionally oriented economic development that results in slow per capita economic growth and technological change. (IPCC 2000)

**Multi-model ensemble**
A collection of results from several different climate models. An ensemble of results allows scientists to investigate the range of uncertainty in the results produced from climate models (i.e., to quantify a degree of model uncertainty).

**Time horizon, *Near-term***
For the purposes of this report, *near-term* refers to a 2010 to 2040 time horizon.

**Non-climate stressors**
For the purposes of this report, non-climate stressors are effects that are unrelated to climate that affect the design, construction, operation, and maintenance of transportation infrastructure.

**Ocean circulation**
The water in the Earth's ocean moves dynamically around the globe, driven by motion of the Earth's atmosphere with the large-scale oceanic circulation pattern (thermohaline circulation) influencing small-scale circulations.

**Ocean salinity**
The concentration of salt within ocean waters, which affects water density and surface water absorption of carbon dioxide.

**Paleoclimate**
The study of past climates on Earth. According to the NASA Goddard Institute of Space Studies, the study of past climates provide insight into how the Earth's atmosphere, oceans, biosphere, and cryosphere has evolved and responded to past climatic forcing (GISS, 2009).

**Precipitation, duration**
A measure describing the length of time of precipitation.

**Precipitation, frequency**
A measure of how often precipitation occurs.

**Precipitation, intensity**
A measure of the rate of precipitation, or the amount of precipitation that falls within a given time period. Typically measured as inches of precipitation per day.

**Precipitation**
In this report, precipitation refers to all forms including rain, snow, or sleet.

**Rain/snow line**
The location within a storm where the precipitation shifts from rain to snow.

**Sea-level, mean**
Average relative sea level over a long enough time period to average out wave/tide variability (such as a month or year).

**Sea-level rise, relative**
The sea level measured by a tide gauge with respect to the land upon which it is situated (IPCC 2007a).

**Sea-level rise, global**
World-wide average rise in sea level. A number of factors contribute to this rise, such as thermal expansion and ice/glacier melting (CCSP 2009).

**Solar radiation**
Wavelike energy emitted by the Sun that possesses heat.

**Spatial resolution**
The level of detail provided by a climate model in assessing climate effects. Spatial resolution refers to the size of the grid used to partition the area being studied. The model calculates climate variable results for each cell within the grid.

**Spring months**
March, April, May (MAM)

**Subsidence**
Local land mass lowers due to plate tectonics. This term is considered when determining relative sea-level rise.

**Summer months**
June, July, August (JJA)

**Thermal expansion (of oceans)**
The increase in molecular motion of ocean water in response to warming, where this motion leads to an increase in volume space for the same number of molecules. Generally considered when estimating how sea level will change as temperatures warm.

**Uncertainty**
Uncertainty in model projections covers three main contributors: natural variability, choice of emission scenarios, and climate models. Uncertainty is further discussed in Section 2.

**Uncertainty; likely range (for the projection of a given climate variable)**
The likely range is computed by first determining the standard deviation above and below the average for each of the two scenarios examined in this report. Assuming the data are well represented by a Gaussian distribution, the likely range represents about 68% of the values extending from the $15^{th}$ percentile to the $85^{th}$ percentile. Then, the minimum and maximum of these four values (i.e., two from each scenario) are defined as the likely range. The range is a measure of the differences (and uncertainty) associated with the models that were used as well as the uncertainty of future emission rates.

**Uncertainty; mean range (for the projection of a given climate variable)**
The mean range is the average of all simulations in the lower emission scenario (B1) and the average of all of the simulations in the higher emission scenario (A2). It is a simple measure of the central tendency of the projections and the uncertainty associated with future greenhouse gas emission rates.

**Uncertainty; very likely range (for the projection of a given climate variable)**
The very likely range is computed the same way as the likely range, except that two standard deviations are used instead of one. Assuming the data are well represented by a Gaussian distribution, the very likely range represents about 95% of the values extending from the $2.2^{nd}$ percentile to the $97.8^{th}$ percentile.

**Uplift**
Local rising of land masses through plate tectonics and/or thermal buoyancy. Uplift is considered when determining local sea-level rise.

**Vertical land motion**
Shifting of land masses through plate tectonics and/or thermal buoyancy. This motion is considered when determining local sea-level rise.

**Winter months**
December, January, February(DJF)

# 6 References

Ahrens, C. 2007. Meteorology Today. Thomson Brooks/Cole, CA, USA. 537pp.

Busey, R., L. Hinzman, J. Cassano, and E. Cassano. 2008. Permafrost Distributions on the Seward Peninsula: Past, Present, and Future. In *Proceedings of the Ninth International Conference on Permafrost* 1: 215-220. University of Alaska, Fairbanks.

California Department of Water Resources (DWR). 2008. Technical Memorandum: Delta Risk Management Strategy (DRMS) Phase 1. 43pp.

Cayan, D., Bromirski, P., Hayhoe, K., Tyree, M., Dettinger, M., and Flick, R. 2006. Projecting Future Sea Level. California Climate Change Center. CEC-500-2005-202-SF. 64pp.

Cayan, D., Bromirski, P., Hayhoe, K., Tyree, M., Dettinger, M., & Flick, R. 2008. Climate change projections of sea level extremes along the California coast. *Climatic Change* 87(0), 57-73. DOI: 10.1007/s10584-007-9376-7.

Cayan, D., M. Tyree, M. Dettinger, H. Hidalgo, T. Das, E. Maurer, P. Bromirski, N. Graham, and R. Flick. 2009. Climate change scenarios and sea level rise estimates for the California 2008 climate change scenarios assessment. California Climate Change Center. CEC-500-2009-014-D, 62pp.

Center for Science in the Earth System (CIG). 2007. Preparing for Climate Change: A Guidebook for Local, Regional, and State Governments. 186pp.

Climate Change Science Program (CCSP). 2007. Scenarios of Greenhouse Gas Emissions and Atmospheric Concentrations (Part A) and Review of Integrated Scenario Development and Application (Part B). A Report by the U.S. Climate Change Science Program and the Subcommittee on Global Change Research [Clarke, L., J. Edmonds, J. Jacoby, H. Pitcher, J. Reilly, R. Richels, E. Parson, V. Burkett, K. Fisher-Vanden, D. Keith, L. Mearns, C. Rosenzweig, M. Webster (Authors)]. Department of Energy, Office of Biological & Environmental Research, Washington, DC., USA. 260pp.

Climate Change Science Program (CCSP). 2008a. Impacts of Climate Change and Variability on Transportation Systems and Infrastructure: Gulf Coast Study, Phase I. A Report by the U.S. Climate Change Science Program and the Subcommittee on Global Change Research. Savonis, M. J., V.R. Burkett, and J.R. Potter (eds.). Washington, DC: Department of Transportation. 445pp.

Climate Change Science Program (CCSP). 2008b. Synthesis and Assessment Product 3.3: Weather and Climate Extremes in a Changing Climate. Regions of Focus: North America, Hawaii, Caribbean, and U.S. Pacific Islands. A Report by the U.S. Climate

Change Science Program and the Subcommittee on Global Change Research. Department of Commerce, NOAA's National Climatic Data Center, Washington, D.C. 164pp.

Climate Change Science Program (CCSP). 2009. Coastal Sensitivity to Sea-Level Rise: A Focus on the Mid-Atlantic Region. A report by the U.S. Climate Change Science Program and the Subcommittee on Global Change Research. [James G. Titus (Coordinating Lead Author), K. Eric Anderson, Donald R. Cahoon, Dean B. Gesch, Stephen K. Gill, Benjamin T. Gutierrez, E. Robert Thieler, and S. Jeffress Williams (Lead Authors)]. U.S. Environmental Protection Agency, Washington D.C., USA. 320pp.

Commonwealth Scientific and Industrial Research Organization (CSIRO). 2006. Infrastructure and climate change risk assessment for Victoria. Report to the Victorian Government. Australia. 177pp.

Department for Transport. 2004. The changing climate: its impact on the Department for Transport. United Kingdom: Department for Transport. http://www.dft.gov.uk/pgr/scienceresearch/key/thechangingclimateitsimpacto1909.

Diffenbaugh, N., J.S. Pal, R. J. Trapp and F. Giorgi. 2005. Fine-scale processes regulate the response of extreme events to global climate change. *Proceedings of the National Academy of Sciences* 102(44):15774-15778.

Ebi, K. and G. Meehl. 2007. The Heat is On: Climate Change & Heatwaves in the Midwest. Pew Center on Global Climate Change. 20pp. Excerpted from the full report: Regional Impacts of Climate Change: Four Case Studies in the United States. http://www.pewclimate.org/docUploads/Regional-Impacts-Midwest.pdf

Frumhoff, P.C., JJ McCarthy, J.M. Melillo, S.C. Moser and D.J. Wuebbles. 2007. Confronting Climate Change in the U.S. Northeast: Science, Impacts and Solutions. Synthesis report of the Northeast Climate Impacts Assessment (NECIA). Cambridge MA: Union of Concerned Scientists. 160pp.

GISS. 2009. Paleoclimate. http://www.giss.nasa.gov/research/paleo/

Grinsted, A., J. C. Moore, and S. Jevrejeva. 2009. Reconstructing sea level from paleo and projected temperatures 200 to 2100AD. *Climate Dynamics* DOI:10.1007/s00382-008-0507-2.

Hawkins, E. and R. Sutton. 2009. The Potential to Narrow Uncertainty in Regional Climate Predictions. *Bulletin of the American Meteorological Society*. pp1095-1107.

Hayhoe, K., C. Wake, B. Anderson, X. Liang, E. Maurer, J. Zhu, J. Bradbury, A. DeGaetano, A. Hertel, and D. Wuebbles. 2008. Regional climate change projections for the Northeast U.S. In press, *Mitigation and Adaptation Strategies for Global Change*.

Hellmann, J., B. Lesht and K. Nadelhoffer. 2007. Chapter 2: Climate. In Hayhoe, K., D. Wuebbles, and the Climate Science Team (eds). Climate Change and Chicago: Projections and Potential Impacts. 33pp.

ICF International (ICF). 2007. The Potential Impacts of Global Sea Level Rise on Transportation Infrastructure. Phase 1 – Final Report: the District of Columbia, Maryland, North Carolina and Virginia. 16pp.

Intergovernmental Panel on Climate Change (IPCC). 2000. IPCC Special Report Emission Scenarios. Summary for Policymakers. 27pp.

IPCC. 2007a. Summary for Policy Makers. In: Climate Change 2007: The Physical Science Basis. Contribution of Working Group I to the Fourth Assessment Report of the Intergovernmental Panel on Climate Change [Solomon, S., D. Qin, M. Manning, Z. Chen, M. Marquis, K.B. Averyt, M. Tignor and H.L. Miller (eds)]. Cambridge University Press, United Kingdom and New York, NY, USA. 996pp.

IPCC. 2007b. Summary for Policymakers, In: Climate Change 2007: Impacts, Adaptation and Vulnerability. Contribution of Working Group II to the Fourth Assessment Report of the Intergovernmental Panel on Climate Change [M...L. Parry, O.F. Canziani, J.P. Palutikof, P.J. van der Linden and C.E.Hanson, Eds.], Cambridge University Press, Cambridge, UK and New York, NY, USA. 976pp.

Kelly, B.P., T. Ainsworth, D.A. Boyce, Jr., E. Hood, P. Murphy, and J. Powell. 2007. Climate Change: Predicted Impacts on Juneau. 86pp.

Kirshen, P., C. Watson, E. Douglas, and A. Gontz. 2008. Appendix: NECIA coastal impact analysis. 4pp. Appendix of Confronting Climate Change in the U.S. Northeast, Science Impacts and Solutions (Frumhoff et al. 2007). .

Knutson, T., J. McBride, J. Chan, K. Emanuel, G. Holland, C. Landsea, I. Held, J. Kossin, A. Srivastava, and M. Sugi. 2010. Tropical Cyclones and Climate Change. *Nature Geoscience* 3:157-163.

Larsen, C.F., R.J. Motyka, J.T. Freymueller, K.A. Echelmeyer, and E.R. Ivins. 2004. Rapid uplift of southern Alaska caused by recent ice loss. *Geophysical Journal International* 158:1118-1133.

Larsen, P., S. Goldsmith, O. Smith, M. Wilson, K. Strzepek, P. Chinowsky, and B. Saylor. 2008. Estimating Future Costs for Alaska Public Infrastructure at Risk from Climate Change. Global Environmental Change, doi:10.1016/j.gloenvcha.2008.03.005. 16pp.

Lenihan, J., D. Bachelet, R. Neilson and R. Drapek. 2008. Simulated response of conterminous United States ecosystems to climate change at different levels of fire suppression, CO2 emission rate and growth response to CO2. *Global and Planetary Change* 64(1-2):16-25

Leung, L., Y. Qian, X. Bian, W. Washington, J. Han and J. Roads. 2004. *Mid-century Ensemble Regional Climate Change Scenarios for the Western United States. Climatic Change* 62:75-113.

Lutgens, F. and E. Tarbuck. 2007. The Atmosphere. Pearson Prentice Hall, NJ, USA. 544pp.

Maurer, E.P., A.W. Wood, J.C. Adam, D.P. Lettenmaier, and B. Nijssen. 2002. A Long-Term Hydrologically-Based Data Set of Land Surface Fluxes and States for the Conterminous United States. *Journal of Climate* 15(22), 3237-3251.

Maurer, E. P., L. Brekke, T. Pruitt, and P. B. Duffy. 2007. 'Fine-resolution climate projections enhance regional climate change impact studies', *EOS Transactions, American Geophysical Union,* 88(47), 504. CMIP3 data is available at: http://gdodcp.ucllnl.org/downscaled_cmip3_projections/

Maurer, E.P. and H.G. Hidalgo. 2008. Utility of daily vs. monthly large-scale climate data: an intercomparison of two statistical downscaling methods. *Hydrology and Earth System Sciences* Vol. 12, 551-563.

Mote, P., A. Petersen, S. Reeder, H. Shipman and L. Whitely Binder. 2008. Sea Level Rise in the Coastal Waters of Washington State. A report by the University of Washington Climate Impacts Group and the Washington Department of Ecology. 11pp.

Mote, P., E. Salathe and C. Peacock. 2005. Scenarios of future climate for the Pacific Northwest. Climate Impacts Group, University of Washington. 13pp.

Mote, P., and E. Salathe. 2009. Future Climate in the Pacific Northwest. Chapter 1 in The Washington Climate Change Impacts Assessment: Evaluating Washington's Future in a Changing Climate, Climate Impacts Group, University of Washington, Seattle, Washington. 23pp.

National Oceanic Atmospheric Administration (NOAA). 2010. Sea levels online. http://tidesandcurrents.noaa.gov/sltrends

National Research Council of the National Academies (NRC). 2008. Potential Impacts of Climate Change on U.S. Transportation. Transportation Research Board Special Report 290. Washington DC: Transportation Research Board. 298pp.

National Science and Technology Council (U.S.). 2008. Scientific assessment of the effects of global change on the United States. Washington, D.C.: National Science and Technology Council (U.S.), Committee on Environment and Natural Resources. 271pp.

New Zealand Climate Change Office (NZCCO). 2004. Coastal Hazards and Climate Change. A Guidance Manual for Local Government in New Zealand. 1st edition. 2nd edition revised by Ramsay, D, and Bell, R. (NIWA). Prepared for Ministry for the Environment. Wellington, New Zealand. 156pp.

Northeast Climate Impacts Assessment (NECIA). 2006. Climate change in the U.S. Northeast: A Report of the Northeast Climate Impacts Assessments. Union of Concerned Scientists, Cambridge, MA. 52pp.

Northeast Climate Impacts Assessment (NECIA). 2008. Climate Change Impacts and Solutions for Pennsylvania: How Today's Actions Shape the State's Future. Union of Concerned Scientists, Cambridge. MA. 62pp.

Ojima, D., J. Lackett, and the Central Great Plains Steering Committee and Assessment Team. 2002. Preparing for a Changing Climate: The Potential Consequences of Climate Variability and Change - Central Great Plains. Report for the US Global Change Research Program. Colorado State University. 103pp.

Payne et al. (2004) as cited in U.S. Climate Change Science Program (USCCSP). 2009. Transportation. In Global Climate Change Impacts in the United States. Second Public Review Draft of the Unified Synthesis Product.

Pfeffer, W.T, J.T. Harper, S. O'Neel. 2008. Kinematic Constraints on Glacier Contributions to 21st-Century Sea-Level Rise. *Science* 321(5894):1340-1343.

Rahmstorf, S. 2007. A Semi-Empirical Approach to Projecting Future Sea-Level Rise. *Science* 315(5810):368-370.

Rohling, E.J., K. Grant, Ch. Hemleben, M. Siddall, B.A.A. Hoogakker, M. Bolshaw, and M. Kucera. 2007. High rates of sea-level rise during the last interglacial period. *Nature Geoscience* 1:38-42.

Salathe, E., R. Leung, Y. Qian, and Y. Zhang. 2009. Regional Climate Model Projections for the State of Washington. Climate Impacts Group. 6pp.

Snover, A.K., L. Whitely Binder, J. Lopez, E. Willmott, J. Kay, D. Howell and J. Simmonds. 2007. Preparing for Climate Change: A Guidebook for Local, Regional and State Governments. In association with and published by ICLEI – Local Governments for Sustainability, Oakland, CA. 186pp.

United States Environmental Protection Agency (USEPA). 2009. US EPA Proceedings: First National Expert and Stakeholder Workshop on Water Infrastructure Sustainability and Adaptation to Climate Change. Arlington, VA. http://www.epa.gov/nrmrl/wswrd/wqm/wrap/workshop.html

Union of Concerned Scientists. 2009a. Confronting Climate Change in the U.S. Midwest: Minnesota. 12pp.
http://www.ucsusa.org/global_warming/science_and_impacts/impacts/climate-change-midwest.html

Union of Concerned Scientists. 2009b. Confronting Climate Change in the U.S. Midwest: Missouri. 12pp.
http://www.ucsusa.org/global_warming/science_and_impacts/impacts/climate-change-midwest.html

Union of Concerned Scientists. 2009c. Confronting Climate Change in the U.S. Midwest: Indiana. 12pp.
http://www.ucsusa.org/global_warming/science_and_impacts/impacts/climate-change-midwest.html

Union of Concerned Scientists. 2009d. Confronting Climate Change in the U.S. Midwest: Wisconsin. 12pp.
http://www.ucsusa.org/global_warming/science_and_impacts/impacts/climate-change-midwest.html

Union of Concerned Scientists. 2009e. Confronting Climate Change in the U.S. Midwest: Ohio. 12pp.
http://www.ucsusa.org/global_warming/science_and_impacts/impacts/climate-change-midwest.html

United States Global Change Research Program (USGCRP). 2000. Climate Change Impacts on the United States, National Synthesis Team, Cambridge University Press. 537pp.

United States Global Change Research Program (USGCRP). 2009. Global Climate Change Impacts in the United States, Thomas R. Karl, Jerry M. Melillo, and Thomas C. Peterson, (eds.). Cambridge University Press. 196pp.

Walsh, J.E., W.L. Chaman, V. Romanovsky, J.H. Christensen, and M. Stendel. 2008: Global climate model performance over Alaska and Greenland. *Journal of Climate* 21(23), 6156-6174.

Wood, A.W., L.R. Leung, V. Sridhar, and D.P. Lettenmaier. 2004. Hydrologic implications of dynamical and statistical approaches to downscaling climate model outputs. *Climatic Change* 62(1-3), 189-216.

Wuebbles, D. and K. Hayhoe. 2004. Climate Change Projections for the United States Midwest. *Mitigation and Adaptation Strategies for Global Change* 9(4):1381-2386.

Yin, J., M. E. Schlesinger and R. J. Stouffer. 2009. Model projections of rapid sea-level rise on the northeast coast of the United States. *Nature Geoscience*, 2(4), 262-266. doi: 10.1038/ngeo462.

Zervas, C. 2001. Sea Level Variations of the United States 1854–1999. NOAA Technical Report NOS CO-OPS 36, 201pp.

# Appendix A. Detailed Methodology

## A.1 Identifying Relevant Climate Change Effects

The climate change effects determined to affect highway systems and discussed in this report include changes in temperature and heat events, changes in precipitation and storm activity, and sea-level rise. These variables are identified as potentially affecting highways by a number of reports (NRC 2008; USGCRP 2009; CSIRO 2006; CIG 2007). Changes in these climate variables may directly affect existing stressors or may introduce new stresses on the highway system. For example, an increased number of heavy precipitation events or the 1-in-100 year storm event (i.e., a storm event with a 1% annual likelihood of occurring) may lead to flooding that cannot be handled by existing culverts and other components of drainage systems. Temperature increases may affect regional highway operations, affecting costs associated with snow and ice removal, as well as the change in environmental impacts associated with salt and chemical use (NRC 2008). Changes in precipitation may disrupt highway travel, construction activities, and compromise bridge integrity (see Section 5 for detailed discussion of climate impacts associated with these climate change effects). Additional variables, such as relative humidity and changes in solar radiation, are also identified but not considered in this report. Literature providing regional information for these variables is currently lacking, but these variables should be included as future research allows.

## A.2 Process of Analyzing Literature and Data

The compilation of regional projections began with a literature search using relevant, carefully composed search terms across relevant databases of publications on the environment, energy, technology, and government. The search was refined to include articles, government reports, and peer-reviewed publications with a published date post-2003 and available by June 2009. Effort was made to include seminal reports that became available after this date. Since this report focuses on regional effects, studies that are broader than the region-scale have been excluded. Sea-level rise studies are an exception, as much of this cutting-edge work pertains to global-scale projections. Any reports that used a data set from the same study are included as a group so that each primary data set is only represented once in the matrix (i.e., not double-counted), thereby avoiding over-emphasis on any one set of projections.

The climate projections provided in the Climate Change Effects Typology Matrix (Appendix C) were culled from studies of varying spatial and temporal scales, modeling parameters, downscaling techniques, and modeling methodologies. This matrix organizes the collected literature by U.S. region, time horizon, climate effect, and spatial coverage with the following characteristics:

- The United States has been divided into regions based on a recent panel-reviewed report (USGCRP 2009);

- The studies are divided into three future time horizons as dictated by the literature findings: as *near-term* represents 2010-2040, *mid-century* represents 2040-2070, and *end-of-century* periods represents 2070-2100;

- The climate variables investigated are listed in the following order: temperature, heat events, precipitation, storm activity, and sea-level rise;

- The USGCRP projection data sets for temperature and precipitation are listed first, followed by the studies with the greatest regional coverage.

We find some results are closely clustered while others range widely, posing a challenge for decision makers attempting to apply them. Attempts to streamline findings are further complicated because many of the studies do not formally quantify likelihood.

## A.3  Description of USGCRP Climate Projections

The bulk of the climate projections used in the regional temperature and precipitation tables in Chapter 3 and used to develop the maps in Appendix B in this report are from the CMIP3 database of climate model integrations by Michael Wehner of the Lawrence Berkeley National Laboratory for use in the 2009 USGCRP report, *Climate Change Impacts on the United States*. The USGCRP data are based on a compilation of aggregated climate model results for two IPCC Special Report Emission Scenarios. These emission scenarios are based on ranges of projections of a number of societal changes, such as changes in population, energy use, technological development, and unpredictable societal behavior (IPCC 2007a; CCSP 2007). The lower emissions (B1) and the higher emissions (A2) emission scenarios in this data set encompass a broad range of possibilities; however, they do not span the full range of possible emissions scenarios. In fact, current emissions rates are exceeding the A2 scenario. On the other hand, emissions scenarios below B2 have been developed for reports of the Intergovernmental Panel on Climate Change, and are being considered in domestic legislation and international negotiations.

This data set is an impressive collection of regional multi-model ensemble results. These values are averaged for each region from the corresponding climate model grid cells in the CMIP3 database. Using GCM-scale resolution is appropriate for this regional projected information of mean conditions of temperature and precipitation.[74] However, the USGCRP report does not provide this information consistently for all regions. We have attempted to be as consistent and quantitative as possible in our documentation of these results to help facilitate comparisons between regions.

---

[74] It should be noted that using the statistical downscaled data of CMIP3 database to provide regional results of mean temperature and precipitation instead of the CMIP3 model results directly would introduce roundoff errors and possibly additional sources of error. Statistically downscaled data are particularly informative when describing fine-scale variability or extremes.

This dataset uses the following global climate models for determining the results of the A2 scenario (moderate-high emissions): ccsm3.0, cgcm3.1, cnrm, csiro, gfdl2.1, hadcm3, hadgem1, inmcm3, ispl, miroc_medres, miub_echo, mpi_echam5, mri_cgcm2_3_2a, and pcm. In addition to these models, the B1 scenario (low emissions) also includes results from: bccr_bcm_2_0, gfdl2.0, giss_aom, iap_fgoals_0_g, miroc_hires. The one exception is for Alaska, where a subset of top-performing global climate models was used: gfdl2.1, mpi_echam5, cnrm, hadcm3, and miroc_medres.

In addition, this report also includes several national maps developed using downscaled data. Statistical downscaling of the CMIP3 database was applied to the larger-scale grid-sized climate models to provide finer-scale results of fine-scaled variability of means and extreme conditions such as heat events that are provided in the figures of this report (Hayhoe et al. 2008; Maurer et al. 2002; Maurer et al. 2008).[75] This downscaling is accomplished by first determining a statistical relationship between surface observations and climate simulations of the past for each region. This statistical relationship is then applied against the results of future climate simulations to provide fine-spatial projections of mean and extreme thresholds of temperature and precipitation. See Wood et al. (2002) for detailed description of this technique. Downscaled data is a useful tool for discussing variability and extremes that are not well captured in climate models (USEPA 2009). It should be noted that using the statistical downscaled data of the CMIP3 database to provide regional results instead of the CMIP3 database itself would introduce roundoff errors and possibly additional sources of errors.

## A.4 Uncertainty

There is always some degree of uncertainty associated with model projections. The Climate Change Effects Typology Matrix includes this information, when possible, in a column labeled "certainty." Modeling the climate system poses a number of challenges, including understanding and representing the climate system's processes and natural variability, and estimating future emissions and uptake of greenhouse gases (IPCC 2007a).

There are various techniques used to address uncertainty, including probabilistic approaches to quantify uncertainty, modeling various emission scenarios to produce a wide range of future possibilities, comparing present-day model results with observations, and engaging expert judgment to express uncertainty based on level of agreement and amount of evidence (IPCC 2007a).

The IPCC assessments (e.g., Fourth Assessment Report (AR4)) and the U.S. Climate Change Science Program (CCSP) Synthesis and Assessment Product (SAP) reports provide some guidance regarding likelihood and confidence and how this information can be used to filter and comprehend projected climate changes. Likelihood represents the

---

[75] Downscaled data is a useful tool for discussing variability and extremes that are not captured well in climate models (USEPA 2009). For access to these downscaled CMIP3 data, see Maurer 2007.

assessed probability that the outcome will occur, and confidence characterizes the consensus across modeling groups or experts that the projections are correct.

Table A-1 outlines the likelihood and confidence for climate variables most relevant to the highway system: temperature rise, changes in precipitation, changes in frequency and intensity of storm events, and sea-level rise.

These likely and very likely indicators provide measures of a portion of the uncertainty and can act as a general guide in assessing the overall findings included in the Climate Change Effects Typology Matrix. However, they do not account for uncertainty associated with future emissions, uncertainty with downscaling techniques, uncertainty associated with the uptake of greenhouse gases, or any systematic errors in the climate models. Hence, the individual studies included in the Climate Change Effects Typology matrix may not reflect the same level of confidence or likelihood as described in Table A-1.

| Climate variable | | Likelihood | Confidence |
|---|---|---|---|
| Temperature Rise | Annual mean | Very likely[a] | High confidence[a] |
| | Seasonal mean | Very likely[a] | High confidence[a] |
| | Extreme Heat Events | Very likely[a] | High confidence[b] |
| Changes in Precipitation | Annual mean | Very likely[a,b] | Not found |
| | Seasonal mean | Very likely[b] | Medium confidence[c] |
| | Change in frequency and intensity | Very likely[b] | Not found |
| Intensification of storm events | | Likely[b] | High confidence (extratropical)[a] |
| Sea-level rise | | Cannot assess likelihood[b] | Not confident in upper bound of SLR[b] |

Table A-1: Likelihood and Confidence for Climate Variables Identified to Affect the Highway System. [a]CCSP (2007); [b]IPCC (2007a); **Very Likely refers to a greater than 90% probability; Likely refers to a greater than 66% probability; High confidence represents an 8 out of 10 chance; Medium confidence represents a 5 out of 10 chance.**

This report quantifies key aspects of the temperature and precipitation projections in the USGCRP dataset. The tables for each region describe the "mean," the "likely" range, and the "very likely" range for each of the three time frames addressed in this study. Those terms are defined as follows:

- o **"Mean"** – The mean range is the average of all of the simulations in the lower emission scenario (B1) and the average of all of the simulations in the higher emission scenario (A2). It is one number where the averages of the simulations for the two scenarios are identical. The mean range is a simple measure of the central tendency of the projections and the uncertainty associated with future greenhouse gas emission rates.

- "**Likely**" – The likely range is computed by first determining the standard deviation for each scenario. Next, the values one standard deviation above and below the average for each scenario are determined. Finally, the minimum and maximum of these four values (i.e., two from each scenario) are defined as the likely range. The range is a measure of the differences (and uncertainty) associated with the models that were used, as well as the uncertainty of future emission rates.

- "**Very Likely**" – The very likely range is computed the same way as the likely range, except that two standard deviations are used instead of one.

Since the studies in the Climate Change Effects Typology Matrix do not necessarily use the same terminology for defining uncertainty, we have limited our use of the terms "likely" and "very likely" in the report's narratives to the aforementioned definitions.

In general, the model projections are more uncertain the further they are into the future. A small range tends to be applicable for the *near-term* time horizon while a larger range of plausible values is appropriate for the long-term. Transportation planners that are less risk-averse may be more comfortable using the "likely" range, whereas planners that are more risk-averse may prefer to use the "very likely" range.

Although the projections described in this report are available at the regional scale, some care should be taken when applying them at regional or local scales. Given modeling efforts currently underway, finer spatial-scale information is likely to become available within the next few years.

## A.5   Comparison of Projections Across Studies

The climate projections provided in the Climate Change Effects Typology Matrix were culled from studies of varying spatial and temporal scales, modeling parameters, downscaling techniques, and modeling methodologies. As a result, some projections are closely clustered while others range widely, posing a challenge for decision makers attempting to apply them. Use of the findings is also complicated because many of the studies do not formally quantify likelihood.

The climate projections contained in the Climate Change Effects Typology Matrix are at a regional-scale or finer spatial resolution. Fine spatial resolution is not directly available from global climate models with a typical grid cell distance varying in size from 50 to 250 miles. Downscaling techniques have been developed that transform the projected climate effect at large-grid cell resolution into a fine-scale resolution of the order of 20 miles or less (NECIA 2006). Such techniques fall into two categories: statistical downscaling and dynamical downscaling. Statistical downscaling determines a relationship between the climate model output of a climate effect for a past 30- to 40-year time period and the observed climate effect for the same time period. This relationship is then used to downscale the projected climate model output for that particular climate effect. This approach is best when the determined relationship is robust over time; that is, the processes governing the climate effect remain fixed with time. This may not always be an appropriate method for downscaling precipitation projections (NECIA 2006). Statistical downscaling is relatively quick and inexpensive. Dynamical downscaling uses a regional model equipped with small-scale processes and local topography. The climate model data are used as inputs around the boundaries of the regional model. Though this process tends to be expensive and time-consuming, it does include dynamical changes in response to large-scale forcing. Given the investment for this approach, dynamical downscaled studies generally provide projections based on the results from a single climate model.

The variation in attributes between studies creates challenges in developing a universal methodology for drafting a regional narrative from the collected projections. Factors that may vary between studies and that can be important when comparing their results include the following:[76]

- **Baseline reference years**: The magnitude or percentage of climate change is computed by comparing model simulations of future conditions with a model's baseline reference period. Therefore, differences in studies' baseline reference periods (e.g., how they define "present day") can introduce unevenness in the amount of change that is reported.

- **Simulation end point**: Comparing the projections of model simulations within the same time horizon can be problematic when the projections are provided for a different set of projected years. This is generally compounded when simulations also refer to varying baseline reference years.

- **Spatial extent of regions**: The spatial extent of a region may be defined differently across studies. For instance, one study may include only coastal locations while another study may also include areas farther inland with very different climate conditions. These differences can directly affect the characteristics of the projected temperature and precipitation changes.

- **Downscaling technique**: The mechanism for downscaling varies between studies. See previous description.

- **Number of global climate models**: Studies vary in the number of global climate models used in the statistical downscaling. For example, the USGCRP data provide averages based on the results from up to 19 global climate models. Other studies, particularly intensive dynamical downscaling studies, may use just one global climate model.

---

[76]Not all modeling assumptions are provided in the "Climate Change Effects Typology Matrix" and therefore can be problematic in efficiently comparing these studies. For example: studies may or may not account for greater degree of ice melting when estimating sea-level rise or studies may or may not allow for evaporation when estimating precipitation amounts.

o **Emissions scenarios**: Most studies we examined tend to draw from four IPCC scenarios: A1Fi, A1B, B1, and A2.[77] They provide results from climate simulations using those scenarios in a variety of ways. For example, some provide the lower and upper limit of the projections determined from climate model ensemble results, while others simply provide the average.

## A.6  Criteria for Data Selection in this Report

It is challenging to synthesize the array of projections provided by the Climate Change Effects Typology Matrix into a narrative discussion that can assist in informing future analysis of climate impacts on the highway system. The criteria used in this report for determining whether a study would be included in the regional narratives of section 4 were guided by discussions with climate experts and are illustrated in Figure A-1. There was a strong consensus for providing a plausible range of projections as opposed to a point value. This range can be provided by studies using multi-model ensembles and relevant emission scenarios (the B2 and A2 scenarios were suggested as the "lower" and "higher" emission scenarios).[78] Further, while a projection associated with a single emission scenario for a multi-model ensemble study is acceptable for *near-term* discussions, as there is not much variation of greenhouse gas emissions between emission scenarios over this time frame, in this study we chose to use two emissions scenarios for each of the three time frames.[79] The projections are, however, affected by the choice of emission scenarios for mid- to long-term projections, as the greenhouse gas emissions increasingly diverge between emission scenarios with time and should be provided across the range of low to high emission scenarios. Providing these ranges can arm transportation planners with a set of plausible scenarios to apply, as warranted, in determining the impacts on the highway system.

The USGCRP (2009) data meet the criteria established for incorporating a study into the narrative: the study was conducted recently, it is a multi-ensemble study, it provides a

---

[77] The IPCC developed four emission scenarios (A1, A2, B1, and B2) described in the IPCC Special Report on Emissions Scenarios (SRES) associated with four plausible storylines representing varying degrees of economic, regional, and environmental projected change as well as allowing for global integration. These studies tend to draw from the following 4 IPCC SRES: **A1Fi**: very rapid economic growth based on per capita, global population peaking in 2050, rapid introduction of new and more efficient technology being fossil-intensive; **A1B**: very rapid economic growth based on per capita, global population peaking in 2050, rapid introduction of new and more efficient technology being evenly distributed between fossil and non-fossil technology; **B1**: rapid changes in economy though slower growth than the A1 scenarios, same global population pattern as in A1, with new technology becoming clean and resource-efficient; **A2**: slowest economic development of all the scenarios based on per-capita growth, has the highest global population allowing for a continuous increase, with the slowest and most fragmented development.

[78] It should be noted that this assumption assumes a somewhat linear relationship between greenhouse gas emissions and changes in the magnitude of the climate variables; hence, it does not account for complexities such as tipping points.

[79] The issue of teasing out climate variability is particularly problematic in the *near-term*. For example, in the 2020s, the regional precipitation projection for the Pacific Northwest was found to lie within the range of natural climate variability, while regional projected temperature was outside of this natural range. The far-term projected temperature variable was found to be far outside the natural range (Mote et al. 2005). This highlights the difficulty in distinguishing between natural climate variability and the projected change.

range of plausible scenarios based on low to moderately high emission scenarios, and the dataset has been aggregated for each region using the CMIP3 database (see section A.3 for further description). In addition, this dataset is uniformly available for each region and time horizon. Therefore, the USGCRP dataset is the primary source for annual temperature, seasonal temperature, and seasonal precipitation. Annual precipitation is not discussed in the narratives, as it averages out the important seasonal variations. One exception of the application of the USGCRP dataset is Alaska. The Alaska projections are based on the five climate models that are determined to be the top performers in simulating temperature and precipitation (Walsh et al. 2008).

Any additional regional temperature and precipitation projections that meet the model inclusion criteria are compared against the USGCRP projections. If the projections are significantly different, the second set of information is provided. It should be noted that this criterion for inclusion into the narrative tends to favor statistical downscaling results over the intensive dynamical downscaling results.

Extreme events such as heat events, and changes in precipitation intensity, duration, and frequency are provided regardless of how well a study meets the criteria discussed above. These studies may use dynamical downscaling followed by multiple regional model runs or intensive statistical downscaling.

Sea-level rise and storm surge projections are not provided by the USGCRP data set. Global sea-level rise studies are discussed at the beginning of Section 4 and are applicable across all regions. Studies that account for regional-scale or local-scale changes in sea level are discussed within the respective region.

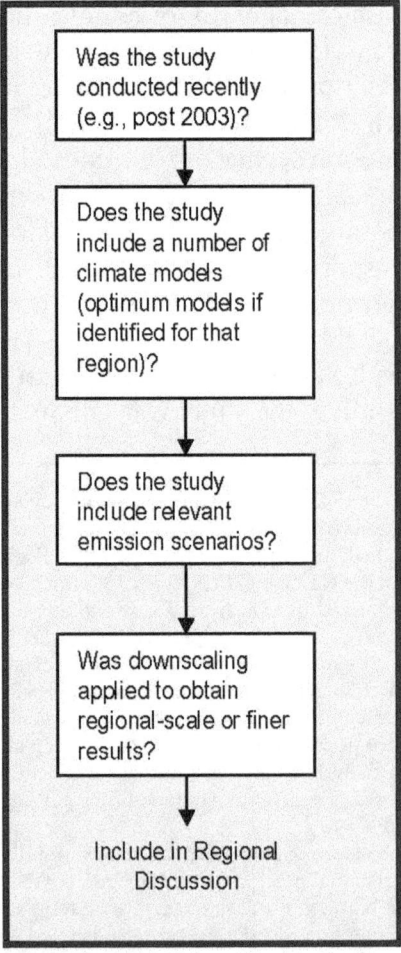

Figure A-1. Criteria for inclusion of study into narrative regional descriptions.

## Appendix B. Regional Maps

*Maps of Temperature and Precipitation by Region*

Attached in a separate document.

## Appendix C. Climate Change Effects Typology Matrix

*Typology Matrix Table by Region*

Attached in a separate document.

www.ingramcontent.com/pod-product-compliance
Lightning Source LLC
Chambersburg PA
CBHW080306180526
45167CB00006B/2688